.e Editors

Dr. Ashok D. Kadlag is Ph. D in Agricultural Chemisty through sandwich programme of IARI, New Delhi and Mahatma Phule Krishi Vidyapeeth, Rahuri. His area of interest is plant nutrition and pesticide chemistry. He is teaching Agricultural Chemistry since 1986. For past 29 years he is engaged in teaching the course of Agricultural Chemistry, Agril. Chemicals and Soil Sciences. He has 53 research paper appeared in national d international journals, 18 books and 78 extension publications. He is develope ˙ ʼ˜ ferent production technologies for different crops grown in Maharashtra. He has guidea M. Sc. and 7 Ph. D. students. He is always keen to update knowledge of post graduate d Ph. D. students and made efforts to help the students to understand basic concepts. He ıwarded as Agrocare Best Teacher in 2013. Presently he is working as a Soil Chemist, Soil st Crop Response Correlation Studies, MPKV, Rahuri (M.S.).

Prof. Arun B. Bhosale is a post graduate in Horticulture from Mahatma Phule Krishi Vidyapeeth, Rahuri. He is working as Assistant Professor of Horticulture at Dryland Agricultural Research Station, Solapur. Since last six year he is engaged in development of production technology for dryland fruit crops. He has developed 8 production technologies of different fruit crops. He has published six research articles in national and international journal, ›ooks and 12 extension papers. He has received Pune Harit Puraskar in 2006.

Sangram D. Kale is a post graduate in Agricultural Chemistry from Mahatma Phule Krishi Vidyapeeth, Rahuri (M. S.). He has ten years of experience in agricultural research and extension out of which five years are at Dryland Agricultural Research Station, Solapur. His area of interest is plant nutrition. He is involved in development of four safflower varieties and sixteen production technologies of fertilizer use for different crops. He has research paper appeared in national and international journals, 10 book chapters and 37 tension publications. Presently he is working as Senior Research Assistant Soil Test Crop sponse Correlation Studies, MPKV, Rahuri (M.S.).

Sanjay M. Todmal is a post graduate in Soil Science from Mahatma Phule Krishi Vidyapeeth, Rahuri. He has ten years of experience in agricultural research and extension. His area of interest is soil fertility management. He is involved in development eight production technologies of fertilizer use for different crops. He has 18 research paper appeared in national and international journals, 6 book chapters and 55 extension publications. esently he is working as Senior Research Assistant Soil Test Crop Response Correlation ıdies, MPKV, Rahuri.

Dryland Fruit Crops

Dryland Fruit Crops

— *Editors* —

A.D. Kadlag

A.B. Bhosale

S.D. Kale

S.M. Todmal

2016

Daya Publishing House®

A Division of

Astral International Pvt. Ltd.

New Delhi – 110 002

Cataloging in Publication Data--DK
Courtesy: D.K. Agencies (P) Ltd. <docinfo@dkagencies.com>

Dryland fruit crops / editors, A.D. Kadlag, A.B. Bhosale, S.D. Kale, S.M. Todmal.
pages cm
Includes index.

ISBN 978-93-5130-972-7 (International Edition)

1. Fruit--Arid regions. 2. Fruit--India. 3. Fruit-culture--Arid regions. 4. Fruit-culture--India. 5. Fruit--Processing--Arid regions. 6. Fruit--Processing--India. I. Kadlag, A. D. (Ashok D.), editor.

SB354.6.A75D79 2016 DDC 634.09154 23

Published by : **Daya Publishing House®**
A Division of
Astral International Pvt. Ltd.
– ISO 9001:2008 Certified Company –
4760-61/23, Ansari Road, Darya Ganj
New Delhi-110 002
Ph. 011-43549197, 23278134
E-mail: info@astralint.com
Website: www.astralint.com

Laser Typesetting : **Classic Computer Services,** Delhi - 110 035

Printed at : **Replika Press Pvt. Ltd.**

Preface

India with diverse soil and climate comprising several agro-ecological regions provides ample opportunity to grow a variety of horticultural crops which form a significant part of total agricultural produce in the country consist of fruits, vegetables, root and tuber crops, flowers and other ornamentals, medicinal and aromatic plants, spices, condiments, plantation crops and mushrooms. India has been bestowed with wide range of climate and physio-geographical conditions, for growing various kinds of horticultural crops. It is estimated that all the horticulture crops put together cover nearby 11.6 million hectares areas with an annual production of 91 million tones. Though, these crops occupy hardly 8 per cent of cropped area in India with approximately 30 per cent contribution in agricultural GDP.

Horticultural crops play a unique role in India's economy by improving the income of rural people. Cultivation of these crops is labour intensive and as such they generate lot of employment opportunities for the rural population.

Fruits and vegetables are also rich source of vitamins, minerals, proteins, carbohydrates etc. which are essential in human nutrition. Hence, these are referred to as protective foods and assumed great importance in nutritional security of the people. Thus, cultivation of horticultural crops plays a vital role in prosperity of a nation and is directly linked with the health and happiness of the people. However, per capita consumption of fruits and vegetables in India is only around 46 kg and 130 g against a minimum of about 92 g and 300 g respectively recommended by Indian Council of Medical Research and National Institute of Nutrition, Hyderabad.

The horticulture sector in India is characterized by small segregated farms with low per hectare yields and huge post harvest losses, owing to outdated practices. A recent techniques of production and processing is needed in horticultural crops.

Horticulturists will have to plays a significant role in the climate change scenario and proper strategies have to be envisaged for saving horticulture. The most effective way is to adopt conservation agriculture; using renewable energy, forest and water conservation, reforestation etc. to sustain the productivity modification of present horticultural practices and greater use of green house technology are some of the solutions to minimize the effect of climate change. Development of new cultivars of horticultural crops tolerant to high temperature, resistant to pests and diseases, short duration and producing good yield under stress conditions, as well as adoption of hi-tech horticulture and judicious management of land use resources will be the main strategies to meet these challenges.

The information compiled were on various aspects of dryland fruit production under climate change scenario would be useful to the farmers research workers, students teachers and policy makers engaged in or associated with soil and water conservation and enhancement in production in dryland agriculture.

A.D. Kadlag

A.B. Bhosale

S.D. Kale

S.M. Todmal

Contents

Preface *v*

1. Introduction **1**

2. Impact of Climate Change in Horticultural Crops **3**
A.D. Kadlag, S.D. Kale and S.M. Todmal

3. Ber **15**
A.B. Pawar, A.D. Kadlag, A.B. Bhosale, S.D. Kale and S.A. Ranpise

4. Pomegranate **57**
S.M. Todmal, S.D. Kale, A.B. Pawar and A.D. Kadlag

5. Aonla **89**
S.M. Todmal, S.D. Kale and A.D. Kadlag

6. Custard Apple **111**
S.D. Kale, S.M. Todmal and A.D. Kadlag

7. Tamarind **131**
A.B. Bhosale, S.D. Kale, S.M. Todmal and A.D. Kadlag

8. Fig **143**
A.B. Bhosale, S.M. Todmal and S.D. Kale

9. Karonda **159**
A.B. Bhosale , R.A. Durgude and S.D. Kale

10. Jamun **181**
A.B. Bhosale, B.D. Tamboli and S.M. Todmal

Index **189**

1
Introduction

Fruits and Vegetables plays an important role in the human diet. It has been reported that there is a considerable gap between the per capita consumption of fruits and vegetables and the actual requirements. Per capita requirements are 255 g. While the actual consumption is about 80 g.

Considering the requirements of fruits and vegetables according to nutritional standards, there is a vast scope, not only for increasing the productivity of the existing orchards but also for increasing area under their cultivation. The present area and production under fruits and vegetables are too inadequate and hence, prices of fruits and vegetables are comparatively higher in the market. This leads to a wrong belief that fruits and vegetables are luxury items of food. To erase this belief, a massive programme of increasing the productivity of existing fruits and vegetable gardens and to exploit all the possibilities of extending their cultivation in suitable areas, will have to be undertaken.

Dryland horticulture has also considerable scope in Maharashtra, cultivation of certain fruits and vegetables crops can be restored in the vast rainfed areas of the state. Fruit crops such as annonas, amla, phalsa, ber, karwand, jambhul, tamarind etc. can be cultivated in dryland/rainfed areas. Since, fruits are highly perishable and can not be stored for a long time; heavy production is always associated with a sudden depression in price. As such grower has no price guarantee. It is therefore, suggested that productivity of fruits and vegetables should be associated with the development of agro based industries which will not only take care of market surplus but will also ensure against sudden fall in market prices.

Maharashtra state has varied agro-climatic zones. It is an ideal state for production of tropical and subtropical fruits. Therefore, it is suggested that marketing and export organizations existing in the country should be strengthened, so that surplus production of high quality fruits could be easily find a way in the competitive marketing channels of other countries.

Mango (*Mangifera indica* L.) is grown as a rainfed crop grown on large area in Kokan region of Maharashtra in regular orchards. Because of the fine quality of fruit, there is a great demand for the Alphonso mango as a fresh fruit and for the processed products such as canned slices and pulps within the country and for export in foreign countries.

Cashewnut (*Anacardium occidentale* L.) is another rainfed crop grown on the hilly regioin of Konkan region of Maharashtra state. The climatic and soil conditions of the Konkan region are well suited for the cultivation of this crop. The area under cashewnut can be extended on the cultivable hilly waste lands of the Konkan region. The constraints of cashewnut is the average yield of cashewnut tree is about 1.5 to 2.0 kg only. Therefore, in general, cashewnut cultivation is not economical. This is on account of the fact that the present cashew plantation are raised from mixed seeds of unknown progeny. Therefore, most of the plants are poor yields. The cashew tree is treated as a wild tree and manure is not used. This is one of the reason for poor yield. The tea-mosquito is a serious pest which causes severe damage to the cashew panicles resulting in very poor yields.

Maharashtra is the leading state both in acrage and production so far as pomegranate is concerned although the total area under the crop in India is 107000 hectares. Out of this 82000 ha area in Maharashtra under pomegranate is almost restricted to the region of state like Ahmednagar, Solapur and Nashik districts. Moreover, it performs well in calcareous soils where other irrigated fruits crops cannot be grown easily. Common cultivars grown in the state are ganesh, arakta, bhagva and super bhagava. Maharashtra sells pomegranate to other states in the country. There is high potential for pomegranates. Hence, there is need to develop a superior type of pomegranate culivars coupled with quality. A package of practices for maximization of yields is being standarised.

Considering the crop requirements for ideal growth it is felt that cultivation of the crop can be extended under the command areas of new irrigation projects such as Sangli, Satara, Pune, Osmanabad, Bhir districts etc.

As the Annonaceous fruits are highly perishable, their cultivation is restricted around the cities. Although the fruit is delicious and posses high nutritional value, the fruits are not so popular on account of the fact that they contain large number of seeds. With a view to evolving a variety of custard apple with less seeds and higher pulp is need to taken on hand addition area under this crop.

In India, Maharashtra is the leading state in commercial cultivation of fig. In Maharashtra the cultivation of fig is restricted to certain pockets of Pune and Aurangabad districts. It is observed that successful cultivation of quality fig is restricted areas where rains are scanty in the months of May to July, because although, some of the districts of the state possess congenial climatic conditions, for the cultivation of figs, the growers are not inclined towards its cultivation for want of irrigation facilities in summer.

2

Impact of Climate Change in Horticultural Crops

A.D. Kadlag, S.D. Kale and S.M. Todmal

Department of Soil Science and Agricultural Chemistry, MPKV, Rahuri, Maharashtra

India and Maharashtra particular with diverse soil and climate comprising different agro-ecological regions provides planty opportunity to grow a variety of horticultural crops which form significant part of total agricultural produce in the country, comprising of fruits, vegetables, root and tuber crops, flowers and other ornamentals, medicinal and aromatic plants, spices, condiments, plantation crops and mushrooms. Horticultural crops plays a unique role in Indias economy by improving the income of the rural people. Fruits and vegetables are also rich source of vitamins, minerals, proteins, carbohydrates etc. which are essential in human nutrition. Thus, cultivation of horticultural crops plays vital role in the prosperity of a nation and is directly linked with the health and happiness of the people. India with more than 28.2 million tonnes of fruits and 66 million tonnes of vegetables is the second largest producer of fruits and vegetables in the world next to Brazil and Chiana. However, per capita consumption of fruits and vegetables in India is only around 46 g and 130 g against minimum of about 92 g and 300 g respectively.

There are about 150 species of human consumable minor fruits grown in India (Mazumdar, 2004). Minor fruits are generally less popular, less palatable, less demanded and less exploited. They are by nature hardy, withstand adverse soil and climate condition and have long protective and productive life. A number of excellent products are made from many of these fruits many are rich source of vitamins, minerals, enzymes, proteins, fats and antioxidants (Das and Das, 2003).

Most of them have medicinal, aromatic and timber and some are utilized as root stocks or in breeding programme. Majority of the minor fruits are naturally less prone to attack by disease due to the least requirement for agro-inputs. Now, the horticulture sector contributes 29.65 per cent of agricultural GDP that has achieved the growth rate of 5-6 per cent during the decade. Food and livelihood security of increasing population is accuse of concern and has received focused attention across the globe. Fruits and vegetables and to food basket, while floriculture and plantation crops enhances the access to food as it provides better income per unit of land besides providing opportunity for employment. Production and export of flowers have increased manifold and the country has a major stake in global trade spices and cashew nuts. Export of medicinal plants, fruits, vegetables have also exhibited rising trend. Admitted, the sector is moving dynamic despite numerous challenges and shortcomings and is a crucial phase of development needing initiatives and sustainable development.

The productivity is subjected to number of stresses and potential yields are seldom achieved with stress. The present challenges like global climate change, water and soil pollution, less water availability, urbanization etc. adds up to the situation. These lead to growing interests among scientists on abiotic stress research. Abiotic stress includes drought, heat, flood, salinity, mineral deficiency, toxicity and chilling or freezing stress. All crops grown under natural environment are subjected to one or the other stress. Horticultural crops require specialized care for high production and quality. Fruits, vegetables, flowers, medicinal plants, tubers are grown from tropical to temperate, some horticultural crops like spices and plantation are location specific. In order to sustain horticultural production with present day challenges one should have packages to manage abiotic stress. The nature and magnitude may vary. Climate change poses serious challenges to human and place unpredicted pressure on the sustainability of horticulture industry. Therefore, the development of horticultural crops that can withstand stress will be the single most important step we may take to adapt the changes we have faced today and will be face in future.

Impact of climate change on four sectors of the ecosystem *viz.* Agriculture, Water, Natural Ecosystems and Biodiversity and Health in four climate sensitive regions of India, namely the Himalayan region, the Western Ghats, the Coastal area and the North East region. The present challenges like global climate change, water and soil pollution, less water availability, urbanization etc., adds up to the situation. In combination with elevated temperatures decreased precipitation could cause reduction in availability of irrigation water and increase in evapotranspiration, leading to server crop water stress conditions. Vegetable production is threatened by increasing soil salinity (Datta, 2013).

Most of the vegetable crops are sensitive to flooding and genetic variation with respect to this character is limited. Tomato crop under flooded condition accumulate endogenous ethylene precursor, 1-aminocyclopropane – 1 – carboxylic acid (ACC) in the roots. During last 4-5 decades air pollution level increasing at an alarming rate in the developing countries and causing potential threat to the crop production. Sulphur dioxide, nitrogen oxide, hydrofluride, ozone and acid rains are the primary air pollutant. Ozone pollution has adverse effect on vegetable production in terms of reducing growth, yield and quality. A recent study indicated that ambient air pollution significantly decreased the yield upto more than 50 per cent. Many vegetable crops namely tomato, watermelon, potato, squash, peas, carrot, beet, turnip etc. are more susceptible to air pollution damage.

Implication of Climate Change in Horticulture

Climate change has implication on two major parameters on agriculture in general and horticulture in particular are *i.e.,* 1) increasing temperature and 2) change in rainfall (both in terms of quantity and intensity).

Climate Change Impact on Horticulture

1. A rise in a temperature of above 1°C will have shifted a major area of potential suitable zones.
2. Production timing will change. Because of rise in temperature, crop will develop more rapidly and mature earlier *e.g.* Citrus, grapes, melons etc. will mature earlier by about 15 days.
3. Photosensitive crops like onion will mature faster leading to small bulb size. Strawberries will have more runners at the expense of fruits by rises in temperature.
4. The faster maturity and higher temperature induced ripening and reduced the storage period in tree/plants.
5. Higher temperature adversely affected the pollination, floral abortions will occur.
6. Annual irrigation requirement will be increased.
7. Reduce the tuber initiation process in potato, reduced quality of tomatoes and poor pollinations in many crops.

Effect on Fruit Crops

India is the second largest producer of fruits after China, with the production of 44.04 million tonnes of fruits from an area of 3.72 million hectares. A large variety of fruits are grown in India, of which mango, banana, citrus, guava, grape, pineapple and apple are the major ones. The horticultural crops will develop more rapidly and mature earlier due to rise in temperature. Delay in monsoon, dry spells of rains and ultimately rains during water stress period, supra optimal temperatures during flowering and fruit growth, hailstorms are some of the most commonly occurring climatic conditions. The climate change increase the atmospheric temperature and change of rainfall pattern, as a result banana cultivation may suffer from high temperature, soil moisture stress or flooding/water logging. Air pollution also significantly reduced the yield of several horticultural crops and increase the intensity of certain physiological disorders like black tip of mango which is induced by coal fume gases, sulphur dioxide, ethylene, carbon monoxide and fluoride. Higher temperature increases the rate of plant maturity in annual species, thus shortening the growth stages. Lower temperature causes the impedance of inflorescence and malformation of bunches.

The plantation crops like coconut, arecanut and cocoa, being perennial in nature face abiotic stress arising due to the climatic change during their life. Since, most of the cultivation part lies in sensitive areas such as coastal belts, hilly areas with high rainfall and high humidity with long dry spell. Impact of climate change on coconut indicated general warming trends in most of the coconut growing areas. Coconut productivity increasing over past fifty years except recent declining trend in Maiden Karnataka, Coimbatore of Tamil Nadu due to consecutive drought. Apart from drought other natural calamities like cyclone etc. have impacted the crop production and productivity.

The abiotic stress arises due to change in climate effects the several morphological, anatomical, physiological, biological and biochemical parameter of the plants. In coconut, arecanut and prolonged drought (> 30 days) significantly reduce the yield and starch content of cassava/sweet potato tubers.

Climate Change Effect on

1. Mango

- ☆ Sudden rise in temperature (maximum 35°C) during peak flowering reduce fruit setting.
- ☆ Low temperature (4-11°C) in January, accompanied by high humidity (> 80 per cent) and cloudy weather may delay panicle emergence.

2. Citrus

- ☆ Winter rains likely to promote vegetative growth instead of flowering flushes.
- ☆ Fluctuations in temperature may affect the bahar management, flower retention, fruit retention and fruit development.

3. Pomegranate

- ☆ Major effect of increased temperatures will be noticed on lower fruit set, fruit drop, fruit cracking.
- ☆ Adverse effect on fruit development and colour development.

4. Banana

- ☆ Increased temperature with low Relative humidity (< 60 per cent) reduced leaf spot disease in banana especially in sub-humid regions.
- ☆ Increased leaf injuries (tearing) because of blowing of hot wind during summer months may result in poor plant growth.
- ☆ Lodging and breaking of pseudostem may occur due to high wind velocity.
- ☆ Poor bunch development and early maturity due to higher temperatures.
- ☆ The banana leaves are more upright under cool condition and more horizontal under warmer conditions.

- ☆ The temperature below 10°C leads to impendence of inflorescence and malformations of bunches.
- ☆ The lamina turns yellow and on the emerging leaf, especially in plant crops, the most recently emerged part of the midrib may show brown areas which are water soaked underneath.
- ☆ Untimely rains and changes in temperature and relative humidity increased thrips and tinged bugs infestation and increases in fungal diseases.
- ☆ Chilling injury during lower temperatures below 10°C which cause bunch chocking may occur.

5. Grapes

- ☆ The wide variation in day and night temperature with high variation (15-20°C) promoting development of anthrocyanin and promoting colour development.
- ☆ Higher temperature affect the growth of the vine in general and fruit bud differentiation in particular after April pruning.
- ☆ Bunch development also affected after October pruning.
- ☆ Pink berry is a major syndrome which deteriorate quality of ruling variety Thompson seedless and this syndrome may increase due to diverse temperatures.

6. Apple and other temperate fruits

- ☆ Change in chilling requirement affected the growth of apple.
- ☆ Higher temperature and moisture stress will cause sunburn and cracking in apple, apricot, cherry and litchi (Kumar and Kumar, 2007).
- ☆ Loosing their hitherto a unique taste and flavour.
- ☆ Specific chilling requirements of pome and stone fruits will be affected hence dormancy breaking will be earlier.

7. Cashew

- ☆ Bright sunshine hours less than 2 hour per day triggers incidence of inflorescence blight in cashew.
- ☆ The tea mosquito bug population and damage score in cashew are negatively correlated with minimum temperature, RH (85-90 per cent) and positively with bright sunshine hours (9-10 hrs.)

Table 2.1: Influence of Weather Parameters on Cashew Flowering and Yield

Sl.No.	*Temperature*	*RH (per cent) A/N*	*Dew Factor*	*Flowering*	*Yield*
1.	15-20°C (min)	> 20	Moderate	Optimum	Satisfactory
2.	21-22°C (min)	> 20	Poor	Reduced flowering	Low yield
3.	> 34.4°C (max)	< 20	High	Drying of flowers	Low yield due to Tea Mosquito Bug (TMB) and flower drying

8. Coconut

- ☆ Impact of climate change related events like consecutive droughts and cyclones adversely affected yields.
- ☆ Elevated CO_2 increased growth of coconut seedling by ≈ 20-30 per cent.

Vegetables

1. Solanaceous Vegetables (Chilli, Brinjal, Tomato)

- ☆ Flower drop, less fruit damage or not fruit set
- ☆ Enlargement of calyx in tomato
- ☆ Errect growth of the plant
- ☆ Reduction in leaf size and more incidence of viral diseases
- ☆ Preanthesis temperature stress is associated with developmental changes in the anthers, particularly irregularities in the epidermis and endothecium, lack of opening of the stromium and poor pollen formation (Sato *et al.*, 2002).
- ☆ Flooded crops especially in tomato plants accumulate endogenous ethylene that causes damage to the plants (Drew, 1979).

2. Cucurbitaceous Vegetables (Cucumber, Ridge Gourd, Bottle Gourd, Muskmelon, Watermelon etc.)

- ☆ More number of male flowers
- ☆ Deformed fruits
- ☆ Bitterness in fruits particularly in cucumber and ridge gourd and more incidence of viral diseases may occur.

3. Bulb Vegetables

- ☆ Temperature fluctuation increase in pre mature bolting
- ☆ Reduction in bulb size
- ☆ Increase in pest and disease incidence
- ☆ Vegetative and bulb development is hampered and reduction in bulb size.

4. Tropical Tuber Vegetable Crops

- ☆ Reduction in marketable grade tuber yield to the extent of 10-20 per cent due to high temperature in potato.
- ☆ Advancement in appearance of aphids by two weeks with increase in 1°C, reduced growing period for seed crop in potato.
- ☆ Elevated CO_2 (550 ppm) enhanced tomato, onion and cassava production and enhanced the quality of sweet potato tubers.

5. Spices

- ☆ Increase in maximum and minimum day temperature and decreasing the annual rainfall, the productivity of black pepper was decreasing in black pepper growing areas of India.

- Seed spices are winter season crops and commonly grown in arid and semi arid track of Rajasthan and Gujarat requiring certain period of low temperature for optimum vegetative growth.
- Combined effect of chilling and frost injury cause heavy losses in yield of spices.
- Cumin, coriander, nigella, ajowan are the crops which are very sensitive to frost.
- Incidence of frost causing serious loss in yield almost reaches upto zero.

Table 2.2: Trends in Climate and Black Pepper Yield at different Black Pepper Production Centers

Sl.No.	Place	Rainfall	T max	T min	Productivity
	KERALA				
1.	Ambalavayal (Wynad Dist.)	Decreasing	Increasing	Increasing	Decreasing
2.	Pampadumpara (Idukki Dist.)	Decreasing	–	Increasing	Increasing
3.	Calicut	Decreasing	–	–	Decreasing
4.	Panniyur (Cannanore Dist.)	Decreasing	Increasing	Increasing	Decreasing
5.	Trichur	Decreasing	Decreasing	No trend	Decreasing
	TAMIL NADU				
6.	Valparai (Coimbatore Dist.)	No trend	No trend	Decreasing	No change
	KARNATAKA				
7.	Kodagu (Coorg)	Decreasing	Increasing	No trend	Decreasing

Source: www.niam.res.in.

Flower Crops

- The impact of climate change on flowering plants and crops.
- Melting of ice cap in the Himalayan regions will reduce chilling required for the flowering of many of the ornamental plants like Rhododendron, Orchid, Tulipa, Alstromera, Magnolia, Saussurea, Impatiens, Narcissus etc.
- Some of them will fail to bloom or flower with less abundance while others will threatended.
- Indigenous species in the natural habitat will be under threat for not getting favourable agro-climatic conditions for their proliferation.
- Western Ghats and surrounding regions may be deprived of normal precipitation due to abnormal mansoon.
- Plant species requiring high humidity and water find difficult condition for survival.

- ☆ Plains of India have similar kinds of problems and will be affected either by drought or excessive rains, floods and seasonal variations.
- ☆ Commercial production of flowers particularly grown under open field conditions will be severely affected leading to poor flowering, improper floral development and colour besides reduction in flower size and short blooming period.

Table 2.3: Soil and Climate Requirement of Horticultural Crops

Sl.No.	*Crop*	*Soil pH*	*Temperature (°C)*		*Rainfall (mm)/ Water Requirement (mm ha^{-1})*
			Min	*Max*	
1.	Mango	5 to 7.5	21	27	750 mm
2.	Banana	5.5 to 8.0	15	35	300 to 700 mm
3.	Grape	5.0 to 8.5	25	35	300 to 750 mm
4.	Papaya	6.0 to 7.5	5	38	300 to 750 mm
5.	Pomegranate	6.0 to 8.5	25	35	300 to 750 mm
6.	Sapota	5.5 to 8.0	11	34	500 to 800 mm
7.	Annona	5.5 to 8.0	8	44	600 to 800 mm
8.	Guava	5.0 to 6.0	11	34	600 to 800 mm
9.	Jackfruit	6.0 to 7.5	25	35	1000 to 1500 mm
10.	Tomato	5.5 to 7.0	15	24	330 mm ha^{-1}
11.	Brinjal	5.5 to 6.0	21	27	486 mm ha^{-1}
12.	Chilli	6.5	20	35	680 mm ha^{-1}
13.	Melons	6.0 to 7.0	30	35	450 mm ha^{-1}
14.	Peas	5.5 to 6.0	10	18	240 mm ha^{-1}
15.	French bean	5.5 to 6.0	15	25	300-350 mm ha^{-1}
16.	Onion	6.5 to 8.0	20	25	500 mm ha^{-1}
17.	Chrysanthemum	6.5 to 7.0	16	25	750 to 800 mm
18.	Jasmine	6.5	21	32	750 to 800 mm
19.	Black pepper	5.0 to 6.0	10	40	1500 to 3000 mm
20.	Cardamom	4.2 to 6.8	10	35	1500 to 5000 mm
21.	Ginger	5.0 to 7.0	19	28	1500 to 3000 mm
22.	Turmeric	4.3 to 7.5	18	30	640 to 4000 mm

Source: www.niam.res.in.

Impact of Climate Change on Water

- ☆ Increasing demand for food, water scarcity is going to be serious problem in near future.
- ☆ Impact of climate change, erratic rainfall and reduced available water are very critical.

- ☆ Water are not managed properly, it will be the single major limiting factor in agriculture sector.
- ☆ Adoption/Development of technologies to improve the water productivity through existing water resources.
- ☆ It could be addressed through judicious use of water and minimizing the losses in conveyance and application.

Pollutants

- ☆ Air pollution causes various types of leaf damage, stunting of growth, decreased size and yield of fruits and destruction of flowers.
- ☆ Plant species sensitive to specific pollutants are useful in monitoring air quality.

Impact of Sulphur Dioxide (SO_2) Pollution

- ☆ Higher level of sulphur dioxide (100-10000 $\mu g\ g^{-1}$) affects leaf drop, leaf curling, reduction in leaf size, premature aging.
- ☆ Inhibits cellular respiration in plants.
- ☆ Increased consumption of sulphur dioxide (SO_2) collapses the plants leaf area and impact on plant productivity.
- ☆ Bleaching of leaf pigments
- ☆ Affects stomatal pores and chlorophyll structure.
- ☆ The leaf area, leaf biomass, productivity, yield etc. are considerably reduced in horticultural crops.

Impact of Ozone (O_3) Pollution

- ☆ Increased ozone concentration near the earth surface reduces crop yield significantly.
- ☆ At 0.02 $\mu g\ g^{-1}$ ozone concentration damages tobacco, tomato, bean, pine and other plants.
- ☆ Ozone alone and in combination with other pollutants like SO_2 and NO_2 is causing crop loss of over 50 per cent in European countries.
- ☆ Ozone affects potato, clover, alfa-alfa etc.

Remedial Measures for Impact of Climate Change

- ☆ Development of tolerant varieties
- ☆ Improve land use management
- ☆ Recycle water and solid wastes in agriculture
- ☆ Change in planting/sowing dates and crop varieties
- ☆ Scientific water and nutrient management

- ☆ Site specific nutrient management
- ☆ Residue management
- ☆ Restoration of degraded soils
- ☆ Conservation agriculture
- ☆ Water harvesting and efficient use of rain water and irrigation water
- ☆ Development of resource conservation technologies for different agro climatic zones.
- ☆ Use of slow release fertilizer
- ☆ Use of nitrification inhibitors
- ☆ Use weather forecasting to avoid rainfall losses.
- ☆ Encourage deeper rooting of crops
- ☆ Improve soil structure to improve water retention
- ☆ Rain water harvesting
- ☆ Re use waste water from farm buildings

In view of these problems, climate change scenario and proper strategies have to be envisaged for saving horticulture. The most effective way is to adopt conservation agriculture, using renewable energy, forest and water conservation, reforestation etc. to sustain the productivity modification of present horticultural practices and greater use of green house technology are some of the solution to minimize the effect of climate change. Development of new cultivars of horticultural crops tolerant to high temperature, resistant to pests and diseases, short duration and producing good yield under stress conditions, as well as adoption of hi-tech horticulture and judicious management of land resources will be main strategies to meet these challenge.

References

Mazumdar, B.C. 2004. *Minor fruit crops of India : Tropical and sub tropical.* Daya Publishing House, New Delhi. P. 145.

Das, B.C. and Das, S.N. 2003. *Cultivation of Minor Fruits*, Kalyani, Publishers, Kolkatta, P. 138.

Datta, S. 2013. Impact of climate change in Indian Horticulture – A Review. *International Journal of Science, Environment and Technology*. 2 (4) : 661 – 671.

Kumar, R. and Kumar, K. K. 2007. Managing physiological disorders in litchi. *Indian Horticulture*. 52 (1) : 22-24.

Sato Suguru, Mary M. Peet and Judith F. Thomas. 2002. Determining critical pre and post - anthesis periods and physiological processes in *Lycopersicon esculentum*

Mill. exposed to moderately elevated temperatures. *Journal of Experimental Botany*, 53 (371) : 1187-1195.

Drew, M.C. 1979. Plant response to anaerobic conditions in soil and solution culture. *Current Advances in Plant Science*. 36 : 1- 14.

www.niam.res.in

3

Ber

A.B. Pawar, A.D. Kadlag, A.B. Bhosale, S.D. Kale and S.A. Ranpise

Department of Soil Science and Agricultural Chemistry, MPKV, Rahuri, Maharashtra

Ber (*Ziziphus mauritiana* Lamk) is an ancient and indigenous fruit of India, China and Malaysia region. The fruits are very nutritious contains considerable amount of carbohydrate and are rich in vitamins–C, A and B complex. The ber is one of the most common fruit trees of India and is cultivated practically all over the country. Ber fruit is within the reach of the poor people and hence known as poor man's fruit.

Ber is much branched, thorny shrub species mostly seen in the Indian desert with a height of 1-2 meter and light coloured bark. It occupies almost all the habitat except the saline patch and sand dunes. The fruits are mostly round berries with black skinned dark brown coloured (CAZRI, 1981). The fruits ripen during late November-December and are much relished.

Among the fruit trees, ber cultivation requires the least inputs and care. It gives good production even without irrigation and can be grown as a rainfed crop in semi-arid and arid region. The Ber tree gives assured income even under marginal growing conditions and provides nutritious food at very low cost. The fruit is dried and used as a desert fruit and it can also be preserved as candied fruit (CAZRI, 1981).

Botany and Ecology

Ziziphus mauritiana is a tropical fruit trees species belonging to the family *Rhamnaceae*. It is a spiny, evergreen shrubs or small tree up to 15 meter height, with 40 cm or more in diameter, priding crown, stipular spines and many dropping branches.

The fruit may be oval, obovate, ablong or round and that can be 2.5-6.5 cm long, depending on the variety. The flesh is white and crispy, when slightly unripe, this fruits is a bit juice and has a pleasant aroma. The fruits skin is smooth, glossy, thin but light. The fresh fruit has mild sub acid flavour and crisp from flesh. It can also be eaten boiled as an additional to rice or millet, stewed or barked.

Medium sized tree that grows vigorously and has a rapidly developing tap root, a necessary adaptation to drought condition. The species widely in height from a bushy shrubs (1.5-2.0 meter) to tall tree (10-12 meter) with a tank diameter of about 30 cm. Ber tree may be erect or wide spreading, with gracefully drooping thorny branches.

The flowers are tiny, yellow, 5 petalled and are usually in two/three in leaf axils. Flowers are white or greenish white and the fruits are orange to brown, 2-3 cm long with edible white pulp surrounding, two locular pyrene.

This quick growing tree starts producing fruits within three years. The fruit is a soft juicy, drupe that is 2.5 cm diameter. The form may be oval, obviate, round or oblong, the skin is smooth or rough, glossy thin but tough. The fruit ripen at different time even on a single tree. Fruit is first green, turning yellow as they ripen. The fully matured fruit is entirely red, soft, juicy with wrinkled skin and has pleasant aroma. The ripe fruit is sweet and sour in taste. Both flesh texture and taste are reminiscent of apples. When under ripe the flesh is white and crispy, acid to sub-acid to sweet taste. Fully ripe fruits are less crispy and some what merely over ripe fruits are wrinkled, the flesh buff-coloured, soft spongy and musky. At first the aroma is apple like and pleasant but it becomes peculiarly musky when over ripe. There is a single, hard oval or oblate, rough central stone which contains two elliptic brown seed, 0.25 inch (6 mm long).

Scientific Classification

Kingdom	Plantae
Division	Magnoliophyta
Class	Magnoliopsida
Order	Rosales
Family	*Rhamnaceae*
Genus	*Ziziphus*
Species	*Mauritiana*
Botanical name	*Ziziphus mauritiana* Lamk.

Ecology

Ziziphus mauritiana is hardy tree that copes with extreme temperature and thrives under dry conditions with an annual rainfall 150-2250 mm. In Fiji, some times naturalized ber trees grow along road sides and in agricultural land, usually near sea level but occasionally up an elevation of about 600 m. It also grows well on laterite, medium black soils with good drainage or sandy, gravely, alluvial soils of dry river beds where it is vigorously spontaneous. In Australia, this species grow on a wide variety of soil types, including cracking clays, sodic soils and deep alluvial in the tropics and subtropics where the average rainfall in the range of 470-1200 mm. The tree is remarkably in its ability to tolerate water logging as well as drought and can grow where annual rainfall ranges from 125-2225 mm, but is more wide spread in areas with an annual rainfall of 300-500 mm. In China and India, wild trees are found up to elevation 5400 ft. (1650 m). In India, minimum shade temperature for survival is 7-13°C and maximum surviving temperature is 50°C. However, this hardy fruit trees not tolerant to frost. Studies report that this species flourishes in alkaline soils with pH as high as 9.2. However, deep sandy loam to loamy soils with neutral to slightly alkaline pH is considered optimum for growth. In India the tree grows on sandy loam, neutral or slightly alkaline. Commercial cultivation of ber usually extended up to 1000 m. Beyond this elevation trees do not perform well and cultivation becomes less economical. A Ber fruit tree does not tolerate frost.

Table 3.1: Ecological Background of Three *Ziziphus* Species

Particulars	*Z. mauritiana*	*Z. jujuba*	*Z. spina-christi*
Latitude	30° N to 30°S	30° S to 51°N	0° to 20°N
Altitude	< 1500 m	Up to 2800 m	< 1000 m
Eco-region	Warm lowland plains	Cool highlands	Mediterranean drylands
Min.Temp.	4° to 12°C	–10° to -20°C	–5 to 2°C
Max.Temp.	39° to 45°C	36°C	±5 ° C
Rainfall	> 300 mm	200 to 450 mm	ca.100 mm
Soil type	Shallow to deep aridisols	Alluvial plains and hills	Poor soils of arid areas
Soil salinity	Neutral to slightly alkaline (5 dSm^{-1})	Highly tolerant	Medium tolerance
Alkalinity	< 45 ESP	Highly tolerant	Some tolerance

Climate

The ber is hardy fruit and grows well all over the country under varying climatic conditions and up to an elevation of 1000 meters above sea level, it favours a hot and dry climate. Ber crop withstand high temperature and aridity by cessation of growth, leaf fall and dormancy phase. In high humidity conditions disease-pest problems increases especially of powdery mildew which is serious disease (Goyal *et al.*, 2008).

In China and India, wild trees are found up to an elevation of 5400 ft. (1650 m), but commercial cultivation extended only up to 3280 ft. (1000 m). In Northern Florida, it is sensitive to frost; young trees may be frozen to the ground but will recover. In India, minimum shade temperature for survival is 44.6 to 55.4° F (7-13°C), the maximum 98.6 to 118° F (37-48°C). The tree requires a fairly dry climate with an annual rainfall of 6 to 88.5 inch (15-225 cm), being unsuited to the lower wetter parts of Malasiya. For high fruit production, the tree needs full Sun (Morton, 1987).

According to Singh *et al.* (1973,b), ber is found growing under varying climatic conditions, almost throughout India. However, temperature below freezing is injurious to the fruits as well as to the young plants. In north-western parts of India during the year, 1974, ber fruits suffered a heavy loss to unprecendented frost. The two species *Ziziphus mauritiana* and *Ziziphus jujuba* are unlike in their climatic requirements. The *Ziziphus mauritiana*, trees goes into dormancy during May-June when it is very hot in the northern plains of India. This period is considered to be the best time for pruning of the plants. The growth, flowering and maturity seems to be more or less correlated with temperature and thus, vary accordingly in different regions. The fruits start ripening as early as middle of December in the western part of India. Whereas, in the northern plains the fruits starts ripening by mid of February.

The drought prone area in Maharashtra covers 114 districts and 1/3 of the total area of the State. The entire Solapur district is covered under scarcity zone. The scarcity zone is traditionally known as the famine tract of the State. Inadequate, erratic, ill-distributed and uncertain with high intensity rainfall are the features of the scarcity zone. The soils are varying in depth from shallow (< 22.5 cm) to deep (> 90 cm). The rainfall ranges from 550-725 mm per year. The climate of the district is hot. The average maximum and minimum temperature ranged between 45 - 10°C.

Temperature

Pareek (1983) outlined the temperature dependency for time of growth, flowering and fruit maturity. Ber trees can withstand extremely high summer temperatures and are found to grow well in regions having maximum temperatures of 39 to 42° C and can tolerate temperatures as high as 49-50° C. However, fruit set is adversely affected at temperatures above 35° C. The trees shed leaves and enter dormancy during the extremely hot summers in the subtropics of northern India. In northern Australia, leaf senescence and leaf fall begins in April as the dry season sets in and the new leaves emerge with the onset of rainfall in September and October (Grice, 1998). This appears to be an adaptive mechanism to escape damage through desiccation during hot weather. Desiccation damage in ber is caused by high cuticular transpiration, which is reported to be associated with high levels of fatty acids and low levels of aldehydes and alcohols in the wax cuticle rather than with the thickness of the cuticle and wax itself (Rao *et al.*, 1981). During the dormancy phase, along with leaf fall, growth ceases. Dark tan-coloured hard and pubescent scales develop on the buds and the bark and buds lose moisture and show lower nitrogen contents (Singh *et al.*, 1974 b).

Rainfall

Natural groves of ber grow and produce well in subtropical and tropical regions receiving over 400 mm annual rainfall. Fruit yields are higher during higher rainfall years (Singh *et al.*, 1998). Management of trees in low rainfall areas necessitates some irrigation. Conditions of high atmospheric vapour pressure deficit result in the production of better quality (sweet and large sized) fruits than in humid areas. The incidence of pests and diseases is also lower in drier regions. Ber exhibits drought tolerance by osmotic adjustment and the deep tap root (Arndt *et al.*, 2000). Species used as rootstocks tend to be those occurring in areas receiving low rainfall. Ber can, depending on the cultivar, withstand slightly waterlogged conditions.

Soil

Ber plant grows on a wide variety of soil ranging from shallow to deep and from gravely and sandy to clayey. The ber develops a deep tap root system within a short period of its growth and as such to adverse soil condition. Ber can with stand alkalinity and slightly waterlogged conditions. It can, in fact, withstand and often do better than most fruits on poor soil.

In India, it grows best on sandy loam, neutral or slightly alkaline. It also grows well on laterite, medium black soils with good drainage or sandy, gravely, alluvial soils of dry river beds, where it is vigorously spontaneous. Even moderately saline soils are tolerated. The tree is remarkable in its ability to tolerate water logged as well as drought.

Ber, because of its hardy nature and tap root system, can be grown in a wide range of soil including porous and unfertile soils unsuitable for major fruit and other crops. Central Soil Salinity Research Institute, Karnal have revealed that ber can also be grown satisfactorily in alkaline soils, characterized by high pH and sodicity. However, sandy loam soils, with neutral or slightly alkaline in reaction, attain good growth of ber plants. Accordingly, Dhankar, *et.al.*(1978), the plants once established continue to grow well even at 21 dS m^{-1} salinity. Dahiya and Dhankar (1982) studied the salt tolerance of ber seedling and found adverse effect of salinity on dry shoot weight, plant height, stem diameter, leaf area, root dry weight. Salt content in soil above 14.2 dS m^{-1} decrease in plant height, leaf number, shoot diameter and fresh and dry weight of roots, leaf contents of P, K, Ca, Mg, Zn and Cu with increase in soil ESP from 10 to 50.

Ber is known to survive even in soils having pH as high as 9.2 (Jawanda *et al.*,1980). No leaf scorching in trees growing on alkaline soils was observed while those of mango and guava were seriously damaged (Samra, 1985). Young ber seedlings survived when transplanted in soils up to 66 ESP (Exchangeable Sodium Percentage) in a greenhouse (Singh, *et al.*, 1983 b). Increasing ESP levels, however, reduced growth, decreased Ca and Mn, but increased Na and Fe contents in the leaves. In pot experiments, seed germination and emergence was reduced in 40 ESP soil and a further increase in ESP resulted in reduction in the amount of P, K, Ca, Mg, Zn and Mn (Patil *et al.*, 1981; Mehta, 1982) and increase in the amounts of Na, B and Fe in the plants (Mehta, 1982). Kumar *etal.* (1990) also observed a decrease in leaf micronutrient content with an increase in ESP levels from 5 to 45.3. The total chlorophyll, free amino acid and proline contents increased with increasing sodicity (Pandey *et al.*, 1991). Awasthi *et al.* (1994) reported that 62.5 per cent ber seedlings survived at 60.5 ESP but none of the grafted plants of Umran and Gola survived. Shoot and root growth of Umran ber decreased with increasing ESP (Patil *et al.*, 1981).

Propagation

The ber was commonly propagated by seeds during earlier period. But main disadvantage of this method is of heterozygosity and variability in seedling progeny.

Therefore, propagation of superior varieties by patch budding is recommended. For raising a budded plantation, it is considered the best to sow ber seeds in the field itself at proper distance and use the seedlings thus, for budding *in situ*. For early germination of ber seeds, breaking of endocarp (hard seed coat) gives quick germination.

Storage of the ripened ber seed for four months improves germination. The hard stone restrict germination and cracking the shell or extraction of seed hastens germination. Without pretreatment the seeds normally germinate within six weeks, whereas, extracted seeds need only one week to germinate.

Seed Propagation

Seed Characteristics

Ber seeds are enclosed within a hard woody endocarp known as the stone which is sometimes wrongly referred to as the seed. Each fruit contains one stone embedded in the pulp at the centre of the fruit. The stone can be depulped in many ways: manually by removing the pulp, pounding dried fruits in a mortar, or by running the fruits through a macerator with water and floating off the pulp. Stones vary in shape from round to subovate to ovate with more or less pronounced ridges on the outer surface (ICUC, 2002). Cleaned stones are about 1-2 cm in size.

Seed Viability

Under natural conditions, seeds from fruits that fall and lie on the soil surface may remain viable for up to 12 months. Data currently available do not suggest that a large proportion of seeds remain dormant in the soil for much longer periods. In Australia, Grice (1996) observed that germination rate in *Z. mauritiana* seeds collected from the soil surface declined from a rate of 56 per cent in the control (fresh dehulled seeds) to 31 per cent after six months, then to 20 per cent at 12 months. For seeds that remain buried in the soil (at 2 cm depth) the germination rate declines sharply to 7 per cent after six months. In Tamil Nadu, India, Srimathi *et al.* (2002) found that *cv.* Umran seeds from fruits collected from the crown exhibited higher germination rates than those collected from the ground.

Storage conditions have an expected significant effect on seed germination. A review by Pareek (2001) indicated that storage at reduced temperatures of 4.5 ± 0.5°C in perforated polythene bags result in retention of viability for longer periods. Ber seeds can remain viable for two and half years when kept in a dry and cool environment, but storage time is dependent on the condition of the seeds and how well they have been dried prior to storage (ICUC, 2002). For long-term storage seeds behave in an orthodox manner and can be stored in low temperatures when dried properly.

A simple and rapid method to test seed viability is to float the stones in salt solution. Seeds that float probably have air pockets caused by insect damage or dead embryos. Such seeds should be discarded. Seeds that sink are viable.

Srimathi *et al.* (2002) soaked dried stones of ber *cv.* Umran in 15, 18 and 20 per cent salt solution and found that only the seeds which sink could be used for seedling production in a nursery. The recovery of floaters increased with increasing salt concentration and germination was improved by grading the stones using 20 per cent salt solution.

Seed Germination

Seed germination is affected by the initial percentage viability at the time of seed collection and by storage conditions, environmental conditions at sowing time, and treatments applied to break dormancy. The degree of fruit ripeness has a significant effect on germination of the enclosed seed. Ber seed germination in Senegal increased from a mere 2 per cent for seeds from green fruits to 28 per cent when fruits are more mature and have turned yellow and 56 per cent for seeds from fully ripe fruits that have turned red. The rate declines to 46 per cent for seeds from overripe fruits that have turned brown (Danthu *et al.*, 1992). This has practical implications for seed collection as fruit colour can be used to determine the most appropriate time to collect seeds.

Pareek (2001) also noted the importance of seed age for optimum germination as ber seeds can germinate only after a period of after-ripening. One or two months after extraction, germination increases and one year old seeds germinate better than the freshly extracted ones. Sowing freshly extracted seeds gave only 33 per cent germination which increased after two months, reaching 50 per cent after storage for eight months. Seedling vigour was also the highest when the seeds were sown after eight months storage. Seed size (small, medium and large) also affects germination. Singh *et al.* (2004) reported the greatest ber seed germination and the tallest seedlings from medium sized seeds.

Seed Pretreatment

Even when they are viable ber seeds do not germinate readily. Suitable conditions for germination must be created. Under natural conditions in the field, seed dormancy gradually breaks down as the stone weathers naturally on the ground. For controlled germination, a common nursery practice is to pretreat seeds if samples take more than a week to germinate. Pretreatment falls into two main categories: mechanical scarification and non mechanical methods.

Mechanical scarification of ber seeds involves cracking, nicking, or partial removal of the stony endocarp. This appears to be the single most efficient pretreatment to achieve the highest percentage of germination.

Non mechanical pretreatments involve soaking whole stones or dehulled seeds in water or in various chemical solutions such as sulphuric acid or a range of growth regulators. A review by Pareek (2001) indicated that ABA- like substances are present in the testa of the seed and that germination was reduced as a result of soaking the seeds in 500 ppm abscisic acid solution. Treatment with 5 ppm abscisic acid inhibits germination of Chinese jujube seeds.

Seedlings to be used as root stock can be raised from seed. Several studies indicated that germination can be improved by soaking seeds in sulphuric acid. Germination time can also be shortened to seven days by carefully cracking the endocarp. Ber seedlings do not tolerate transplanting; therefore, the alternatives are to sow the seeds directly in the field or to use polythene bags placed in the nursery bed. Seedlings are ready for budding in 3 to 4 months. In addition, seedlings from the wild cultivars can be converted in to improved cultivars by top working and grafting. Nurseries are used for large scale seedlings multiplication and graft production. The seedlings should also be given full light.

Standardised propagation techniques for ber were established. Budding is the easiest method of vegetative propagation used for improved cultivars. Different types of budding techniques have been utilised with ring budding and shield budding being the most successful. Wild varieties of ber are usually used as the root stock. The most common species used for root stock are *Z. rotundifolia* in India and *Z. spinachristi* in Africa.

These can be propagated by seeds, however, seedlings take several years to start fruit production. Superior varieties can be air layered or grafted. Grafted plants are less thorny than seedlings.

Nursery Management

Seed Propagation

Ber seedlings raised from seeds are not used directly for fruit production in commercial plantations because the time taken for such plants to reach bearing age is usually longer than for trees propagated using vegetative methods (ICUC, 2002). Also the desirable traits of improved cultivars (growth, fruiting and fruit quality) are not guaranteed through seed reproduction. None the less, propagation of ber and the other major jujubes by seed remains of great value for several reasons:

1. Propagation by seed is necessary to raise appropriate rootstocks for vegetative propagation of improved cultivars.
2. In regions where jujubes are propagated not solely for fruits but also for a range of products and services (live hedges, fodder, timber etc.), propagation by seed is a much easier and more economical method than vegetative propagation. Until the recent introduction of improved cultivars to the Sahel region of West Africa, most research and development programmes on *Z.mauritiana* in the region focused on performance in live hedge technology in agroforestry systems under rainfed conditions. This requires large scale production of seedlings sown directly in the field or raised in private and community nurseries in villages throughout the region. In such situations, propagation by seed is the safest way to produce seedlings with deep tap roots that can survive in the field under extended moisture stress even when the surface soil completely dries out.
3. Sexual reproduction by seed offers opportunities for variation and evolutionary advancement, which is of immense value to the plant breeder.
4. Conservation of *Ziziphus* biodiversity and rehabilitation of natural vegetation is the best served through seed reproduction.

Techniques for Raising Seedlings

Seedlings may be raised in nurseries by sowing seeds in seedbeds or in containers. Quality seedlings in the nursery are fundamental to quality trees in the field. As Jaenicke (1999) pointed out, the best looking seedling in the nursery is worthless if it does not survive and grow after planting out.

For ber, seeds should be sown at a depth of 2 cm at 30 cm x 30 cm spacing in the seedbed any time from spring through the monsoon period (Pareek, 2001). For transplanting in the field, nursery bed raised seedlings have to be lifted, packed and transported to the planting site. Seedlings can be transplanted in the field with 20 cm x 15 cm earth ball; they can also be transplanted bare rooted (without ball) when they are not growing actively during the winter in subtropical climates (Pareek, 2001). Bed raised seedlings often suffer high mortalities during transplanting because of root injuries when seedlings are lifted from the bed, or because of insufficient care during packing and transportation to the planting site. Hardening of seedlings can be beneficial before transplanting (Bhatia *et al.*, 2001).

From a review by Pareek (2001), it appears that treatment of roots with 250 ppm IBA improves transplanting success. Shifting the seedlings within the nursery one month after sowing and then transplanting them in the field after months also results in better survival. Nonetheless, overall risks of mortality remain quite high when transplanting bed raised seedlings. Thus, the method of raising seedlings in nursery beds is not very popular. Seedlings raised that way are not considered suitable for planting under rainfed conditions.

Vegetative Propagation

Methods of vegetative propagation differ between Z. *mauritiana* and Z. *jujube.* Traditional methods such as use of cuttings are not successful with the former whereas, in Chinese jujube use of cuttings is, and furthermore Chinese jujube cuttings can be taken from root suckers before emergence. Some success has been observed in ber by stooling. Z. *spinachristi* is propagated by cuttings.

Budding is the major method for propagating ber and can be successfully carried out using several methods. Of the different methods, the degree of success varies according to climatic conditions, time and method of budding and the skills of the operators.

Both ber and Chinese jujube can be propagated by grafting, using the wedge method for ber and the whip or tongue method for Chinese jujube.

Rootstocks

For ber, rootstocks are either the same species (wild) or commonly Z.*nummularia* (often referred to by its synonym Z. *rotundifolia*). No incompatibility is seen between scions of the cultivars and rootstock seedlings of Z. *nummularia,* wild Z. *mauritiana, Z. oenoplia, Z. rugosa, Z. xylocarpa, Z. jujuba* or Z. *spina-christi* (Pareek, 1983). None the less vigour and growth differ greatly. Use of Z. *nummularia* tends to be less favourable for fruit production than the use of Z. *mauritiana* (Pareek and Nath, 1996).

Rootstock seedling plants of different species, although from variable progenies, influence the vigour, growth form and productivity of the scion and presumably also its adaptation to adverse climatic and soil conditions (Verma *et al.*, 2001).

The best choice, at present, seems to be to raise rootstocks from seeds collected from vigorous trees of cultivated ber (Z. *mauritiana*) and boradi (Z. *mauritiana* var. *rotundifolia*) or other suitable wild species such as Z. *abyssinica.* Use of vegetatively propagated or seedling plants from identified genotypes of the selected species should in due course lead to the development of standard rootstocks. This will ensure the desired effects of adaptability, productivity, vigour and form of the scion, and longevity of the scion/rootstock combination under different ecological habitats.

In Chinese jujube, seedlings of Z. *jujuba* var. *spinosa* and other wild species are used as rootstocks (Ming and Sun, 1986). Grafting over 2 cm thick roots of *Paliurus spinosa* has also given success (Zaktreger and Solov'ev, 1962).

Budding

The Budwood

Budwood becomes available during the active growth period in the summer. The bud sticks, with well swollen and recently matured buds (but still not open) are collected. Immature and undeveloped buds from the upper part of the new shoots and over mature and inactive buds should not be used. Buds collected from lowering shoots give poor success. Buds should be collected from juvenile shoots. Such shoots can be induced to grow by severe pruning of the mother trees.

After collection, the budwood is often stored for a time or has to be transported, and considerable loss of viability may take place. Budwood retains good availability when kept under ventilated conditions and wrapped in moist jute cloth. Kanwar and Singh (1981) found that buds stored in polythene sheeting, even for one day at 31-38° C and 67-68 per cent relative humidity, failed in bud-take. However, buds kept in moist jute cloth for one day gave bud-take of 68.8 per cent and 36.2 per cent when kept for five days. Singh (1988) found that budwood wrapped in polythene sheeting and moist sphagnum moss gave better results than those stored in moist jute cloth.

Time of Budding

The best time for successful budding is during the active growth period. The active growth period is indicated by easy and clear separation of the bark from the wood in both scion and the rootstock. Kaundal *et al.* (1984) observed maximum bud-take (80-87 per cent) at ambient temperature between 30 and 34° C, irrespective of variations in atmospheric relative humidity (RH) from 45 to 73.5 per cent. Bud-take declines below 36 per cent RH and when temperatures drop to 18.5-20° C. Such conditions occur during the summer and monsoon period (June to September) in the arid and semi-arid subtropics of northwest India (Joshi, 1960; Singh *et al.*, 1972; Pareek. 1978 a; Singhrot *et al.*, 1980; Anonymous, 1981; Kaundal *et al.*, 1984; Mawani and Singh, 1992 a, b). Maximum sprouting of 93.2 per cent has been recorded by budding when the maximum temperature was between 33 and 36.5° C and minimum temperature was around 26° C (Kaundal *et al.*, 1984). Some success can be obtained even during the mild winter months of the tropics (Jyotishi *et al.*, 1967).

Budding Method

Budding can be carried out by different methods as I or T (shield) or ring and patch budding (Husain, 1973) and (Mawani and Singh, 1992a,b), chip (Anonymous, 1953), flute (Mawani and Singh, 1992, a,b). However, shield and patch budding are the most commonly used methods. The ring method needs the scion stick to be equal diameter to the rootstock and this limits its use on a large scale. Using patch budding 75-85 per cent success has been obtained (Anonymous, 1981). Standardisation of the budding methods has been summarised by Nayak and Sen (2000).

Budding should be performed as close to the ground level as practicable to minimise the area for emergence of sprouts from the rootstock portion. Over 93 per cent success has been obtained by budding at the height of 10 cm onto seedlings with a stem diameter of 0.53 - 0.55 cm (Singh *et al.*, 1981 b; Chattopadhyay and Dey, 1992). Lopping and topping of the rootstock from (Kundu 1983; Singhrot and Kajal, 1986) 0 to 15 days (Singh *et al.*, 1984 b) after budding in the nursery beds has been found to ensure 90 to 100 per cent success. Topping has also been found to increase growth of buddings (Anonymous, 1981). Pareek *et al.* (1999) recommend that just before *in situ* budding in the field, rootstock seedlings should be lopped at 30-45 cm height. When budding onto seedlings raised in polythene bags, the rootstocks should be prepared by removing all side shoots and leaves from the stem up to a height of 15 cm above ground level. The seedlings are then topped just before budding using either the shield or patch method.

Transplanting of Budded Plants

Budded seedlings prepared in nursery beds are lifted with large earth balls at about 9 to 12 months after budding for transplanting to the field. A number of these plants may be lost due to damage while lifting, packaging, during transport and transplanting (Singh *et al.*, 2001). The operations are cumbersome and incur high costs due to the large earth balls. Attempts have been made to reduce mortality by:

1. Shifting the plants within the nursery 25 days after sprouting (6 leaf stage) to harden them before final transplanting at about 8 months after budding (Singhrot and Makhija, 1979 a; Bhatia *et al.*, 2001).
2. Shifting the buddings into polybags of 30 cm x 20 cm size along with treatment of 12 per cent Waxol (Kundi and Singhrot, 1990).
3. Transplanting the defoliated buddings without earth ball during the winter in the subtropics (Sandhu and Dhillon, 1983; Sandhu *et al.*, 1983; Beniwal *et al.*,1992).
4. Cutting the tap roots along with treatment with 200 ppm IBA (Singh, 1988) and wrapping the roots with moist sphagnum moss or jute cloth during transport to the field.

All these methods increase establishment success, but only after irrigation as transplanting tends to reduce the drought-hardy character of ber trees owing to loss of tap roots.

Buddings prepared in polybags become ready for transplanting at about 30 days after budding. The polybags with the buddings are removed from the nursery and kept in shade for a week. These can then easily be transported and transplanted to the field with over 90 per cent survival (Pareek, 1978 b) since there is little stress or root damage during lifting, transport and planting. The roots of the plants do not coil and therefore, retain the drought-hardy character and vigour almost similar to plants raised *in situ*. For long distance transport, soil is washed from the roots of polybags raised buddings and covered with sphagnum moistened with 0.2 g potassium nitrate, 0.8 g calcium nitrate, 0.2 g magnesium sulphate, 0.2 g hypophosphate and 1 ml of 0.5 per cent ferric tartrate with 0.2 per cent Dithane Z-78 (Pareek and Vashishtha, 1980).

Planting

Spots are marked at each site. In India pits of 60 cm^3 size are dug in April or May for planting during the monsoon or December to January for spring planting (Bajwa *et al.*, 1972; Singh *et al.*, 1973 a; Bakhshi and Singh, 1974). In arid areas the best time for planting is at the onset of the monsoon (Pareek, 1983). Pits can be treated to control termites and are filled one month before planting with a mix of topsoil and 15-20 kg farmyard manure. Planting then takes place. In arid areas this is early but in others it can extend over two months.

Planting distance depends on soil types, cultivar, climatic conditions and amount of training to be imposed on the plants. Higher density is possible on poor soils if training is vigorous. Where possible, drip irrigation is provided after planting (Singh, 1973 a) for the first two months. Distances are shown in Table 3.2.

Table 3.2: Planting Distance of Ber

Distance in (m)	*Plants ha^{-1}*
6x6	277
6x5	333
5x5	400
6x4	416
5x4	500
6x3	555
4x4	625

Source: Singh, 1992.

Training of Ber Plants

For developing a strong frame work of the tree, it is essential to train them right from the nursery stage. Training is done during the first three years after planting.

If ber tree is not trained and left to itself, it would attain bushy, large unmanageable form but the production per unit area is quite low. In order to keep the plant in manageable shape and size, trees may be trained properly during first two to three years to built up a strong frame work.

First Year

- ✰ An upright and vigorous main trunk is developed.
- ✰ It is kept clean of secondary branches up to 75 cm height.
- ✰ From the main trunk 4-5 well spaced main branches are allowed to grow.
- ✰ The top is headed back.

Second Year

- ✰ Secondary branches are allowed to grow, keeping a spacing of 60-75 cm in between.
- ✰ Tertiaries are allowed to grow.

Third Year

- ✰ Final balancing and correction of the tree framework is done during the third year.
- ✰ Tertiaries are allowed to grow maintaining a spacing of 60-75 cm between them.
- ✰ Pruning is done in the month of May.

Thus, plants are trained by pruning and kept in manageable shape, size with well developed frame work.

Water Management

Availability of water is the major determinant for growth and productivity of all cultivated jujubes.

Rainfed Areas: Water Harvesting

Many ber orchards are found in dry areas and depend only on annual rainfall for water. Productivity of these orchards depends mainly upon the quantity and distribution of rainfall and the extent of its subsequent use by the trees. A considerable quantity of rain water is lost as runoff and cannot be used by the trees unless water harvesting methods are employed. These may be *in situ* where runoff is concentrated into catchments surrounding a tree or group of trees, or *ex situ* where runoff from large rocky areas is diverted directly to nearby orchards or after collection in reservoirs or farm ponds.

For *in situ* water harvesting land shaping may be used to prepare sloping catchments on both sides of the rows of ber trees. In the subtropical sandy planes, which receive 360 mm annual rainfall in northwest India, higher runoff yield was obtained from catchments having 5 per cent slope than from the usual land slope of 0.5 per cent (Yadav *et al.*, 1980; Sharma *et al.*, 1982, 1986). In due course, the sandy catchments stabilised as a result of compaction and levelling and generated runoff even with rainfall of only 25 mm.

On natural undulating wastelands, trees can be planted at the lowermost point of each microcatchment where runoff accumulates. For this, the area is divided into microcatchment plots of 150 to 500 m^2 each. The size of plots depend upon the required quantity of runoff supplement which in turn depends upon rainfall, catchment slope, antecedent catchment characters (compaction, smoothness, etc.) and the runoff coefficient of the soil (Evenari *et al.*, 1971). Bhati *et al.* (1997) used rainwater collected in small farm ponds to irrigate ber trees during the fruit development period.

Irrigated Areas

Irrigation during early growth can be beneficial. Drip irrigation is most effective for this (Yaragattikar and Itnal, 2003). Farmers do not generally irrigate established ber orchards other than for intercrops (Verma and Gujar, 1994). However, water deficits decrease the productivity of trees by reducing the assimilation rate. In the subtropical semi-arid region of north India, irrigation during the fruit development period has been observed to increase productivity by decreasing fruit drop (Bakhshi and Singh, 1974). Although irrigation during the lowering period causes flower shedding in ber (Singh *et al.*, 1973 a), watering may have to be done if there are prolonged breaks in the monsoon during the growth and flowering period. Irrigation at the time of fruit maturity delays and prolongs the maturity period.

Conservation of Moisture

Mulching conserves moisture in the soil and suppresses weed growth. A range of organic wastes can be used. Normally a bed is prepared around the trunk covering three quarters of the area of the canopy. Black polythene sheeting is also useful and the most effective treatment (Byun *et al.*, 1989, 1991 and Ming and Sun, 1986). A number of chemical sprays have been tested to reduce loss of water through transpiration. Other chemicals can be watered in the soil around the trees for this purpose. However, they are not yet cost-effective.

Weeding

Weeding is normally carried out through regular hoeings which are shallow tillings. Particular attention in all jujube species is paid to removal of root suckers in order to maintain the individual tree. Annual ploughing of inter-row spaces helps weed control when carried out after weed emergence, followed by harrowing later.

Herbicides can be used e.g. Dalapon followed by Paraquat, each at the rate of 5 kg ha^{-1} (Bajwa *et al.*, 1990) and glycophosphate at 4 kg ha^{-1} (Bajwa *et al.*, 1993).

Manures and Fertilisers

Small orchards are seldom manured. In the arid region of northwest India, farmers do not apply any manure or fertiliser in ber orchards other than that given at planting time (Verma and Gujar, 1994). Regular manuring is, however, essential to replenish the nutrient removal by the tree through fruit harvests and annual pruning besides losses from the soil. According to Ahlawat *et al.* (1990), as many as 75 per cent of the orchards in the arid subtropics were deficient in organic N and 90 per cent in P but none in K. Under subtropical and tropical agroclimates, nitrogen deficiency is widespread and thus, it plays a major role in determining fruit yields. Deficiencies of N, P and K reduce shoot length, leaf number and size, axillary branch number, flowering and fruit set and consequently productivity of the tree (Sadhu *et al.*, 1978).

A ber tree removes from the soil, 142 to 191 g N, 59 to 87 g P and 467 to 684 g K on average during a single growing season (Mehrotra *et al.*, 1987). Most is due to fruit harvesting. According to Pandey *et al.* (1990), 43.8 g N, 7.5 g P, 101.2 g K, 4.5 g Ca and 22.7 g Mg are removed by 100 kg harvest of fruits. Judicious manuring and fertilization are necessary to replace nutrient losses and to correct deficiencies due to imbalances in the soil or plant. Nitrogen and phosphorus fertilization in the light soils of the arid and semi-arid subtropics is critical and potash may be applied to guard against its possible limitation as a result of annual removal over several years.

Nutrients Recommended

In irrigated ber orchards 20-30 kg FYM or 400-500 g N, 200 g K_2O and 200 g P_2O_5 can be provided annually when planted widely. In rainfed conditions half of these doses are recommended (Singh, 1992).

The method of placement of fertilisers depends on the root growth pattern. In an irrigated orchard, most of the active roots of 10-year old ber trees are located 2.35 m from the trunk and up to 0.5 m deep (Khera *et al.*, 1981). Under rainfed conditions, however, the roots make more vertical growth down to a depth of 4.5 m (Pareek, 1977).

Like other fruit trees, ber also requires regular application of manures and fertilizers for good yields. Application of 30-50 kg farm yard manure, 250 g N per tree (Split in 2 equal doses) and 250 g P_2O_5 per tree + 50 g K_2O per tree (Single dose) per year for full grown trees (5 year and thereafter) is recommended for better yield and quality of fruits Deficiency of N, P and K was found to inhibit shoot length, leaf number and axillary branches production, the flowering was also reduced and the plants failed to set fruits (Sadhu *et al.*, 1978).

Application of 40-50 kg farm yard manure in July and 0.75 kg ammonium sulphate per tree in October (after fruit set) was recommended by Singh and Singh (1973). Daulta and Chauhan (1982) recommended application of 50 kg well decomposed farm yard manure during dormant period and one kg of nitrogenous fertilizers per tree in two split doses, one during rainy season and again at the time of fruit set. Yamdagni *et al.* (1980) observed that irrespective of the levels (250,500 or 750 g/tree) provided with 500 g N per year showed a yield from 86 to 91 kg per tree as compared with 61 to 67 kg under control (Table 3.3). Kumar and Tung (1981) reported an increase in vigour of *cv.*, Umran by fertilizing the plants with 480 g N, 150 g P_2O_5 and 300 g K_2O annually. Increasing doses of N, P and K also increased the leaf N, P and K contents.

Table 3.3: Effects of Levels of Nitrogen, Phosphorus and Potassium on Yield, Number of Fruits and Stem Girth of Ber (Pooled means of 7 years)

Sl.No.	*Nutrient Dose*	*Yield (kg tree^{-1})*	*No. of Fruits Tree^{-1}*	*Stem Girth (cm)*
	Nitrogen (g tree^{-1})			
1.	0	44.32	2296	84.76
2.	250	47.55	2317	83.40
3.	500	48.19	2234	82.96
4.	750	49.12	2328	82.36
	SE±	0.34	27.33	0.13
	CD 5 per cent	0.95	75.77	0.35
	Phosphorus (g tree^{-1})			
1.	0	44.14	2276	80.64
2.	250	47.96	2303	64.93
3.	500	49.04	2303	84.54
	SE±	0.29	23.67	0.11
	CD 5 per cent	0.82	65.62	0.30
	Potassium (g tree^{-1})			
1.	0	44.72	2260	80.95
2.	50	48.89	2338	84.15
3.	100	48.54	2283	85.01
	SE±	0.29	23.67	0.11
	CD 5 per cent	0.82	65.62	0.30
	Interactions	NS	NS	NS

Source: Anonymous, 2007.

Table 3.4: Ber Yield, Number of Fruits and Stem Girth as Influenced by Interaction between Levels of Nitrogen, Phosphorus with Potassium (Pooled means of 7 years)

Nutrient (g tree^{-1}) N:P_2O_5	*Yield (kg tree^{-1})*			*Number of Fruits Tree^{-1}*			*Stem Girth (cm)*		
	0	*50*	*100*	*0*	*50*	*100*	*0*	*50*	*100*
0:00	45.10	46.62	47.53	2228	2264	2300	79.5	83.9	83.9
0:250	46.99	48.19	49.67	2356	2385	2376	82.5	87.1	86.7
0:500	46.04	47.23	48.33	2234	2232	2291	83.9	88.0	87.4
250:00	44.43	45.11	47.49	2245	2230	2259	78.3	82.6	82.0
250:250	47.01	47.85	49.05	2345	2397	2445	81.6	86.0	85.1
250:500	47.72	48.43	50.90	2253	2302	2376	82.2	86.6	86.3
500:00	46.97	46.48	47.43	2253	2274	2255	77.6	82.2	82.0
500:250	48.39	48.92	49.98	2292	2355	1898	81.3	85.1	85.0

Contd...

Table 3.4–*Contd...*

Nutrient (g tree^{-1}) N:P_2O_5	*Yield (kg tree^{-1})*			*Number of Fruits Tree^{-1}*			*Stem Girth (cm)*		
	0	*50*	*100*	*0*	*50*	*100*	*0*	*50*	*100*
500:500	48.02	48.51	49.01	2204	2243	2329	81.8	86.0	85.7
750:00	47.80	46.94	48.50	2259	2268	2287	77.3	81.4	80.9
750:250	47.81	51.83	50.94	2392	2363	2452	80.5	84.4	84.5
750:500	49.40	49.22	49.62	2250	2320	2363	81.3	86.0	85.0
SE±		1.020			82.01			0.38	
CD 5 per cent		NS			NS			NS	

Source: Anonymous, 2007.

The ber fruit trees are provided with nutrients on annual basis. Nitrogen dose may be supplied through organic and inorganic forms 50 per cent each as follows

Table 3.5: Fertilizer Schedule for Ber

Year	*FYM (kg tree^{-1})*	*Fertilizers (g tree^{-1})*		
		N	*P_2O_5*	*K_2O*
1st year	10	100	50	50
2nd year	20	200	100	100
3rd year	30	300	150	150
4th year	40	400	200	200
5th year	50	500	250	250

DRIS

The Diagnostic Recommendation and Integrated System (DRIS) was employed to interpret nutrient analysis of leaf tissues from ber fruit orchards grown in semi-arid and arid areas of Punjab in north west India. The DRIS indices, which assessed nutrient balance and their orders of limitations to yield (Singh *et al.*, 2010). Nutrient sufficiency ranges derived from DRIS norms are as follows:

N- 0.688 – 1.648 per cent
P- 0.184 – 0.339 per cent
K- 1.178 - 1.855 per cent
Ca- 1.064 – 1.768 per cent
Mg- 0.234 – 0.391 per cent
S- 0.124 – 0.180 per cent
Fe- 55 – 205 mg kg^{-1}
Mn- 26 - 80 mg kg^{-1}
Zn- 17 – 33 mg kg^{-1}
Cu- 5 – 11 mg kg^{-1}

A field experiment was conducted consequently for three years at the Agricultural Research Station, Anantnagar, Andhra Pradesh, India on shallow red sandy loam soil. The recommended dose of fertilizer to ber @ 30 kg FYM tree^{-1}, 100 g N, 50 g P_2O_5 and 50 g K_2O with optimum catchment area per plant gave the higher fruit yield and improved the fruit quality (Reddy and Umamaheswari, 2003).

Foliar Feeding

Under rainfed conditions, foliar feeding is often a practical method of supplementing nutrient requirements, particularly of nitrogen and micronutrients. Nitrogen fertilization of ber trees through foliar sprays of urea increased fruit set, fruit retention and yield and improved fruit quality in cultivars Banarsi Karaka (Rajput and Singh, 1976; 1977), Umran (Chauhan and Gupta, 1985; Singh and Ahlawat, 1995) and Gola (Joon *et al.*, 1984). Dahiya *et al.* (1985) found that application of 375 g N through foliar spray and 375 g N to the soil gave a fruit yield of 128-130 kg per tree, higher than by applying all the nitrogen either through soil or foliar. The optimum concentration of urea for foliar application seems to be 1 to 2 per cent (Joon *et al.*, 1984; Chauhan and Gupta, 1985; Dahiya *et al.*, 1985; Singh and Ahlawat, 1995). Two sprays may be required at monthly intervals depending upon the nitrogen requirement. The best time for spray application seems to be after the fruit has set (Joon *et al.*, 1984; Chauhan and Gupta, 1985; Dahiya *et al.*, 1985; Singh and Ahlawat, 1995).

Supplementing micronutrients through foliar sprays of 0.2 to 0.4 per cent $ZnSO_4$ $FeSO_4$ or $MnSO_4$ and 0.2 per cent H_3BO_3 has been observed to increase TSS and ascorbic acid contents (Kamble and Desai, 1996). This resulted in development of golden-yellow rather than green-yellow fruit colour in tropical regions. In the arid subtropics where zinc deficiency is more common and widespread, higher concentrations of $ZnSO_4$, *i.e.* 0.5 per cent (Singh and Ahlawat, 1995) and 0.8 per cent (Joon *et al.*, 1984) have given a better response.

Combined sprays of micronutrients (0.03 per cent boric acid and 0.5 per cent $ZnSO_4$) and growth regulator NAA (50 ppm) have been found to improve fruit quality in ber by increasing total soluble solids, total sugars and ascorbic acid and decreasing acidity (Singh *et al.*, 1989 a,b).

In Chinese jujube 200 ppm ethephon along with urea sprays accelerated movement of nitrogen from leaves to other parts of the tree and increased the length of fruit bearing branches, fruit set and fruit weight (Hao and Zeng, 1991).

Microbial Inoculations

Microbial inoculations of soil and ber roots with *Azospirillum*, *Azotobacter* (Rao and Dass, 1989), vesicular arbuscular mycorrhizal fungi (VAM) (Mathur and Vyas, 1995 a,b,c,d; 1996) and other nitrogen fixing actinomycetes (Tuohy *et al.*, 1991) can help in effective nutrient management. Soil inoculation with cell suspensions of *Azospirillum brasilense* strains S14, S51 and S54 or *Azotobacter chroococcum* increased height and weight of young plants of *cv.* Gola (Rao and Dass, 1989).

Pruning

Pruning is an essential operation in ber production as fruits are born in the axils of leaves on the young shoots of current season. Pruning is, therefore, done every year to induce maximum number of new healthy shoots which bear good quality fruits. The best time for pruning was observed to be 15th April – 15th May under prevailing climatic conditions of Maharashtra or when it sheds its leaves and becomes dormant which is observed after harvesting of the fruits.

Dry, dead and diseased wood, cris-cross branches and weak crotches should be removed at the time of pruning. After the establishment of the frame work, pruning should be continued every year but drastic thinning and very severe pruning should be avoided, severe pruning adversely affects the growth of plant and gives poor yield. For optimum yield and good quality fruits, 25 per cent of the one year old shoots should be headed back.

Pruning is essential to maintain vigour in the trees and to maintain fruit productivity, quality and size.

Fruit bearing in ber is on the current season's shoots (Singh *et al.*, 1970; Pareek, 1983) and remains confined to the secondary and tertiary branches (Reddy, 1983). Pruning should therefore, induce the emergence of a maximum number of secondaries and tertiaries on vigorous shoots. This can be done by pruning at the right time and with the right intensity depending upon location and cultivar.

Pruning Intensity

The severity of annual pruning is determined by the length of past seasons shoots retained after the pruning operation, example as below:

Length of Last Season Shoot (cm)	*Pruning Intensity*
20 or 25	Very sever
50 or 60	Sever
70 or 75 or 90	Moderate
100 or 120 or 125	Light
150	Very light

Source: Singh *et al.*, 1978b, Gupta and Singh, 1979, Lal and Prasad, 1979, 1980 a, b, c; 1981, Bajwa *et al.*, 1986, 1987 and Nanthakumar, 1991.

However, the severity of pruning varies depending on the vigour of the shoot (total length and diameter). An alternative approach has been to base light, moderate and severe pruning on removal of one-quarter, half and three-quarters of the length of the shoot respectively (Sharma, *et al.*, 1980; Singh and Godara, 1985; Yadav and Godara, 1987, 1992). The three severity levels have also been based on pruning the shoot at a point where it is 1, 2 or 3 cm in diameter (Bisla *et al.*, 1988, 1990, 1991).

Old, unpruned, jujube species can be rehabilitated by reducing overlapping branches and precipitate fruit bearing towards the outer part of the crown. Half the number of shoots on a tree are pruned at 15 to 25 nodes depending upon cultivar, and the remaining half are pruned severely retaining only the basal node (Kundu *et al.*, 1995). From the basal node of the severely pruned shoots, vigorous foundation shoots emerge which during the next year are pruned at 15 or 25 nodes while the remaining half are pruned severely. This pattern of pruning helps to maintain sustainable production as well as tree form. Even 30 year old trees can be brought back to bearing (Bal *et al.*, 2004)

Kurian (1985) was able to induce early bud break and increase the number of vigorous sprouts by a pre-pruning spray of 3 per cent thiourea combined with two post-pruning sprays of 100 ppm benzyladenine or 50 ppm TIBA at monthly intervals. A pre-pruning spray (before 48 h) with 3 per cent thiourea and then pruning half the shoots at 25 nodes and the remaining half to the previous season's growth gave the highest fruit yield of 128 kg/tree.

Pruning Time

Whereas, Chinese jujube is pruned in summer the time of pruning in Indian jujube varies in different climates. In subtropical regions, the most appropriate time is during summer (between May and June in parts of India) when the trees shed their leaves and enter dormancy (Gupta and Godara, 1989; Gupta *et al.*, 1990; Kundu *et al.*, 1995) and before putting forth new growth. Early pruning has been observed to advance flowering and fruit maturity (Singh and Sandhu, 1984; Pareek and Nath, 1996).

In tropical regions with a mild winter and rainfall during December-January and early onset of summer rainfall (*e.g.* Tamil Nadu), pruning can be carried out any time from January to April. It is also possible to regulate fruit maturity so that bearing occurs at the desired time (Pareek and Nath, 1996). In tropical regions without winter rainfall (*e.g.* Andhra Pradesh), pruning during the first two weeks of April results in early flowering and therefore, early harvest; however, the maximum yield of good quality fruits is obtained by pruning in the second two weeks of April (Ramadevi, 1989). In Maharashtra, the best time for pruning is before the end of April (Pareek and Nath, 1996; Deotate *et al.*, 1997) as further delay causes reduction in fruit yield.

Plant Growth Regulators

Plant growth depends on the environment. The growth kinetics are subject to change due to weather parameters *viz.*, solar radiation, temperature, rainfall and relative humidity. Growth of the plant is promoted when conditions are ideal for development. Each fruit crop or species has definite requirement of above parameters and this is why the training, pruning and bahar treatments are different for various fruit crops. Some fruit crops are adapted to hot weather and some requires cool environment for better growth. Flowering of crop is dependent on photo period, hot, dry, humid and cool climate accordingly the fruit crops are categorised as tropical, subtropical, arid, sub-arid etc. fruit crops. Different seasons are characterised by specific photoperiod, temperature and relative humidity. Ideal weather means the initiations of desired changes in hormonal levels that can promote the growth under normal nutritional and water requirements. Variation in soil, water and temperature influence the endogenous hormonal levels and varies the physiological processes affecting the growth and yield.

Dryland fruit crops are subjected to variation in available soil moisture, changes in relative humidity and variation in day and night temperature. Adverse soil and weather conditions brings the changes in endogenous hormonal levels their by influencing the growth. The water stress promotes the level of growth inhibitors and suppressed the level of promoters. The balance of these chemical substances governs the growth either good or bad. The water stress/drought results in production of abscisic acid, ethylene, phenol compounds and oxidants that deteriorates the growth.

Table 3.6: Role of Plant Growth Regulators on Physiological Processes

Sl.No.	Plant Growth Regulators	Physiological Processes
1.	Auxin, IAA, Kinetin, 2,4-D	Cell division, enlargement, differentiation, germination, flowering, mitochondrial activity.
2.	Gibberellins	Parthenocarpy, fertilization, ovul development, germination, pollen viability, seed development, sex expression, dorfism, dormancy.
3.	Cytokinins	Plastid, chloroplast development, determent of senescence, enzyme activity, organogenesis.
4.	Abscisic acid (ABA)	Leaf and fruit abscission, defoliation, embryo genesis, dormancy, stomata closer.
5.	Ethylene (CEPA)	Wilting, fruit ripening and retention, sex expression, ethylene activity, breaking dormancy.
6.	Growth inhibitors (CCC, PCB, Resorcinol, Malic hydrazide)	Dwarfing, branching, source-sink relationship, antigibberellins, inducing dormancy.

Source: Dhopate, 2002.

Use of Plant Growth Regulators in Ber Production

Attempt to increase fruit set, fruit size and early/delay maturity through application of growth regulators proved to effective. The growth regulators like gibberellic acid (GA_3), Naphthalene acetic acid (NAA), Cycocel (CCC) and ethephon were used. GA_3 (10 ppm) and NAA (10 ppm) during fruit development and ethephon one month before harvesting were sprayed. Application of NAA increased the yield of fruits.

Spray application of 20-30 ppm NAA once in the second week of October and again in the second week of November check, 11 per cent fruit drop in Sanur-5 and 10 per cent in Kaithali cultivar of ber. One of the main constraints in fruit production in some cultivars of ber is occurrence of high natural fruit drop (about 70 per cent). It is established that plant growth regulators can significantly change the hormonal status of the plant resulting good fruit retention and their by improve the fruit yield and quality (Ghosh *et al.*, 2009).

Spraying of GA_3 and NAA (10 and 20 ppm) at full bloom and 15 days after flowering increased fruit weight and size appreciably with 20 ppm GA_3. The maximum decrease in stone weight of ber by 10 ppm NAA. Quality of fruit was improved with 20 ppm GA_3. Total soluble solids, reducing sugars was increased and decreased in acidity by 10 ppm NAA and 20 ppm GA_3. (Kale *et al.*, 2000).

The effect of growth substances on fruit set, growth and development of fruit has been studied on ber.

In an investigation, on the effect of NAA, 2,4-D, 2,4,5-T and methyl ester of naphthalene acetic acid (MENA) at 10 – 30 ppm (Singh and Singh, 1976) observed that sprays of 2,4,5-T reduced fruit drop by 14 -19 per cent and increased fruit size and TSS. Two sprays of NAA @ 25 ppm, 1st at after fruit set and 2nd one month of fruit set reduced the fruit drop 9.8 per cent and improved the fruit size and TSS (16.2 per cent) of *cv*. Sanaur-2 (Bal *et al.*, 1982).

The highest fruit set in ber *cv.* Gola was observed in application of NAA at 10 to 30 ppm at the time of flowering and again after 15 days (Bankar and Prasad, 1990).

Yield

Variation in yields of Z. *mauritiana* cultivars (55.5 to 116.11 kg/tree) have been observed by Kumar *et al.* (1986). In the semi-arid subtropical climate of northern India under irrigated conditions, the fruit yield per tree ranges from 80 to 200 kg depending on the varieties and management practices during the prime bearing age of 10 to 20 years (Bakhshi and Singh, 1974). Fruit yield was the highest in Gola (38.4 kg tree^{-1}) and lowest in the local cultivar Sukavani (3.54 kg tree $^{-1}$) at Sardar Krushinagar in Gujarat (Chovatia *et al.*, 1992). The highest yields were obtained from Umran (210 kg/tree) followed by Sanaur No. 2, Dandan and ZG 2 (Gupta, 1977). Thornless gave the highest yield/tree (74.4 kg) followed by Sanaur 5 (71.3 kg) and Sanaur 4 (Singh and Tomar, 1988) at Bhatinda, Punjab, India. Kharki had the highest fruit yield at Hoshangabad, Madhya Pradesh (Gupta and Mehta, 2000).

At Hyderabad, Mundia was the highest yielder (116.1 kg/tree) followed by Umran and Gola (Kumar and Babu, 1987) whereas, yield in different cultivars of Z.*mauritiana* in Maharashtra (India) indicated maximum fruit, pulp and stone weights in Kadaka when pruned on 25 March (Bharad *et al.*, 2002). Banarasi Karaka fruits had the highest nutritive value and yield (99.8 kg fruits/tree) (Tiwari and Banafar, 1995). Reddy *et al.* (1998) carried out an economic analysis of cost and returns to determine the most profitable cultivars at Dharwad, Karnataka, India. The performance of 11 cultivars showed that Dandan, Sanaur-2 and Chhuhara, with mean fruit yields of 6.78, 6.36 and 6.08 t ha^{-1} and benefit: cost ratios of 5.64, 5.32 and 5.09, respectively were the most promising. Cultivar Pewandi is considered the best for commercial cultivation in Uttar Pradesh, India (Teaotia *et al.*, 1974).

Chinese jujube cultivars yield 50 to 300 kg fruits per year depending upon cultivar, location and age of tree (Ciminata, 1996). On the basis of yield, ripening date, frost resistance and storage quality in 7 ber cultivars Ta-Yantszao, Da-bai tszao, Ya-tszao, 93, 58, 107, 52 and 48 are recommended for cultivation (Sin'ko, 1977). In the Arava valley of Israel, a fruit yield of 12 tonnes per hectare has been obtained from 3 year old ber trees (Nerd and Mizrahi, 1998). From the naturally growing, scattered wild trees in Zimbabwe, a yield of 4-5 tonnes has been obtained from a 3-4 hectare area (Maposa and Chisuro, 1998). In China - Taiwan, fruit yields of 158.6 kg per tree in cultivar Kaolang-1 and 140.8 kg per tree in cultivar Telong have been obtained (Chiv Chu Ying, 1997).

Improved Cultivars

Numerous horticultural varieties of ber are grown all over India. Some of the most popular varieties are Umran, Karaka, Gola, Seb, Chhuhara, Sanaur-2, Illachi and Mehrun. In Maharashtra *cv.*, Umran is being grown on commercial scale. It has excellent keeping quality and transportability but flat taste.

Ber cultivars are differing in the habit of the tree, leaf shape, fruit form, size, colour, flavour, keeping quality and fruiting season. In India more than 90 cultivars are cultivated by the ber growers. Among the important cultivars eleven are described in the encyclopaedic "Welth of India" *viz.*, Banarashi (Banarasi), Pewandi, Dandan, Kaithli (Patham), Muria Mahrara, Nankelee, Nazuk, Sanauri-1, Sanauri-5, Thornless and Umran (Umri).The skin of most is smooth and greenish-yellow to yellow.

Seventy cultivars of ber were studied at Haryana Agricultural University. These cultivars were collected from all jujube growing areas of northern India and set out in an experimental orchard in 1967-68. In 1980, sixteen midseason selections from these were evaluated "Banarasi Karaka" (Poor flavoure) gave the highest yield (130 kg per tree) followed by Mudia, Murhara and Kaithali (both of good flavour) and Sanauri-5 and 'Desi Alwar' (both of medium flavour). Mudia, Murhara, Kaithli and Sanauri-5 were worthy of commercial cultivation. For breeding purposes Banarasi, Karaka and Desi Alwar could contribute high pulp content. Mudia, Murhara - total soluble solids, Kaithali - high ascorbic acid content and good flavour, in efforts to develop a superior mid-season cultivar. Ber varieties can be classified on the basis of following criteria:

Depending upon Maturity Period

Early	*Mid Season*	*Late*
Gola	Kaithali	Umran
Safed selection	Sanauri-5	Kathphal
Seb	Reshmi	Ponda

Depending upon Region of Cultivation

Commercially Grown	
Maharashtra	*Other States*
Umran	Kaithali
Kadaka	Kathaphal
Sanur-2 and 6	Ponda
Punjab chhuhara	Muria
Illaichi	Seb
Gola	Safed selection
Mehrun	Jogia

Distribution of Important Traits

The important traits in some ber cultivars have been described by Saran (2005) (Table 3.7) and the traits identified by others are shown in Table 3.8.

Table 3.7 Characteristics of Promising Ber Cultivars

Sl.No.	Cultivar	Size	Colour	TSS (per cent)	Acidity (per cent)	Av. Fruit Weight (g)	Yield (kg/tree)	Other Features
1.	Umran	Large	Golden-yellow turning chocklet brown when fully ripened	19	0.12	30-89	150-200	Late ripening Good keeping and shipping quality
2.	Gola	Medium to Large	Golden-yellow	17-19	0.46- 0.51	14-17	80-100	Juicy of good flavour Earliest to ripen Sells at high price
3.	Kaithali	Medium		18	0.5	180	100-150	Pulp soft and sweet
4.	Kathaphal	Small to medium	Greenish blushed on one cheek with reddish yellow	23	0.77	10	80-100	Late in season
5.	Gular Bashi	Medium	Yellow-brown	18.8 (yellow colour) 22.4 (brown colour)	–	–	–	Juicy sweet, non acrid Excellent quality when fresh Musky after storage Late in season Keeps well
6.	Kheera	Medium to Large		19.8	–	–	–	Pulp soft, juicy Good sweet flavour Fairly good keeping quality Late in season
7.	Nazuk	Medium to Small		17.4	–	10.5		Pulp slimy Fairly juicy Sweet flavour Without stringency Moderate bearer Poor keeping quality
8.	Seo ber (Seb)	Medium to Large	Light brown	19.0	–	–	–	Pulp moderately juicy Stone large Late in season Keeps very well

Table 3.8: Important Traits in Ber

Cultivar	*Characters*
Umran	Yield, stem girth, fruit size, fruit weight, flesh thickness, flowers per cyme and pulp/ stone ratio.
Ponda	Yield, spread, number of flowers per cyme, fruit weight, fruit size and flesh thickness
Sanauri-5	Yield, number of flowers per cyme, fruit weight, size of fruit and flesh thickness
Laddu	Yield, spread and number of flowers per cyme
Gola	Spread, fruit weight, flesh thickness and stone weight
Chhuhara	Spread, stem girth

Important traits have been identified by various workers which can be used by breeders; some breeders are listed in Table 3.9.

Table 3.9: Traits identified in ber cultivars

Traits	*Cultivars*	*Author*
Fruit maturity	Early (Gola, Mundia), mid season (Banarsi, Kaithali), late (Umran)	Vashishtha, 1983
Sweetness (high TSS)	Reshmi, Umran	Vashishtha,1983
Pulp texture	Coconut-like (Umran), Juicy (Gola, Aliganj), Melting (Illaichi)	Vashishtha, 1983
Fruit size	Very large (Ponda), large (Umran), Medium (Mundia, Banarsi, Gola), small (Illaichi)	Vashishtha, 1983
Fruit shape	Apple like (Seb), cardamomshaped (Illaichi), bell shaped (Mundia), Round (Gola), oblong (Umran)	Vashishtha, 1983
Fruit colour at maturity	Bright golden (Sanaur), brightyellow (Gola), Greenish yellowwith brown blush (Kathaphal)	Vashishtha, 1983
Acidity	Very low (Umran), low (Gola),Moderate acidic (Sanaur), acidic (Kathaphal)	Vashishtha, 1983
Shelf life of fruits	Good (Umran, Maharwali), poor (Gola)	
Processing uses	Dehydration (Vikas, Raja, Babu,Jeevan, Chinese cultivars, Umran, Bagwari, Chhuhara) Preserve (Umran, Banarsi Karaka, Kaithli) Candy (Illaichi, Umran, Kathaphal, Kaithli)Beverage (Gola, Mundia)	Gopani, 1976 b, Gupta *et al.,* 1981b, Gupta *et al.,* 1981a Khurdiya and Singh, 1975
Resistance to fruitfly	Tikadi, Meharun, Illaichi	SinghandVashishtha, 1984
Resistance to fruit borer	Banarsi Pewandi, Gola Gurgaon, Jhajjar Selection	Pareek and Vashishtha, 1986
Tolerance to powdery mildew	Illaichi Jhajjar, Sanaur-5, SafedRohtak, Kathaphal, Gola, Seb,Meharun	Pareek and Nath, 1996
Field resistance to powdery mildew	Dharkhi-1, Dharkhi-2, Guli,Villaiti, Seedless	
Tolerance to *Isariopsis*	Safed Rohtak, Sanaur-1, SeoBahadugarhia, Jhajjar Selection	Pareek and Nath, 1996

Source: Pareek, 2001.

Attributes of Important Indian Ber Fruits

Umran

This cultivar is commercially cultivated on a large scale in Punjab and Haryana States of India. It was developed from germplasm from Rajasthan at the Fruit Research Station at Bahadurgarch, Punjab. It fetches the highest price. The fruits are large sized, oval in shape and have a roundish apex. They weigh on average between 30 and 80 g. They are an attractive golden yellow colour which later turns into a chocolate brown at full maturity. The fruit matures in the mid-season (February to March) and ripens during mid-March to mid- April. The fruit is sweet with 19 per cent total soluble solids (TSS) and 0.12 per cent acidity. It has a pleasant flavour and excellent dessert quality. Umran fruits have a good keeping quality and can withstand long transportation. The main reason for its popularity is the long shelf life (15 to 20 days) and excellent organoleptic qualities. The fruit is also known locally as Ketha, Ajmeri and Chamdi.

Kathapal

This is a late ripening cultivar of ber well known in Gujarat that has small to medium size fruits. At maturity, the fruits remain green on one side while the other side develops a reddish yellow tinge. The average fruit weighs 10 g. They have a high TSS (23 per cent) and an acidity of about 0.77 per cent.

Gola

Gola is an early-maturing cultivar that is grown in Uttar Pradesh, Gujarat, Punjab, Rajasthan, Haryana and Delhi. It starts to bear fruit in the first week of January. The fruits are very attractive, roundish in shape and golden yellow in colour with an average weight of 20 g. The white flesh is very juicy, semi-soft and has a delicious taste. The fruit pulp has a TSS of 17 to 19 per cent and 0.46 to 0.51 per cent acidity. The ratio of pulp to stone is 14.

Kaithali

This is a mid-season cultivar of ber well known in the Punjab that ripens during March. It was developed from the collection from the Kaithali area of Karukshestra in Haryana. The fruit is medium in size with an average weight of 18 g, oval in shape and has a tapering apex. The fruit pulp is quite soft, has a TSS of 18 per cent and 0.5 per cent acidity. Unlike Umran, this fruit does not withstand transportation and has a poor keeping quality.

Tikadi

This ber cultivar is a late maturing type that ripens in Rajasthan during February and March. The fruits are small, weighing on average 10 g. The fruits become edible, with a creamy soft flesh, when the skin turns to a red colour during a short (7 to 10 days) ripening period of the fruits on the trees. The ripe fruit has a high TSS (25°) with a large stone (the pulp to stone ratio of ripe fruit is 6.9).

Jogia

The fruits of this cultivar are also well known in Rajasthan and have a light purple tinge when they are unripe, but are still edible at this time. Because of this, they have an extended harvest period up to early February. The skin surface has coarse ridges and is greenish yellow in colour. The flesh is white, soft, juicy and sweet with a TSS of 19° and a pulp to stone ratio of 14.

Mundia

This is an early Rajasthan, high yielding, cultivar that matures during mid-January. The fruits are large, juicy and bell shaped with an average weight of 40 g. They have a yellowish green skin with smooth depressions. The flesh is white and soft with a TSS of 20° and a pulp to stone ratio of 23.

Fruit Based Agroforestry System for Dryland

Integration of fruit crops in crop lands is referred to as agri-horticultural land use system of Agroforestry. Aonla, ber, guava, citrus etc. were identified as promising fruit crops and are the first preference of farmers under agroforestry system on account of short gestation period, regular income, risk cover and aesthetic value. Agri horticulture land use is an important component of agroforestry. Agroforestry is land use that involves deliberate retention, introduction or mixture of trees or other woody perennials in crop/animal production fields to benefit from the resultant ecological and economic interactions.

Agri-horticultural system or horti-pasture system is one form of agroforestry where the tree component is a fruit tree. Agroforestry is a collective for systems of land use in which woody plants (trees or shrubs) are deliberately combined on the same land management unit with herbaceous crops in some form of special arrangement.

Among the alternate land use system developed, dryland horticulture and agri-horticulture system (fruit based agroforestry system) are readily picked up by the farmers due to cash benefits derived from these system.

Ber is considered as the king of arid zone fruits and performed well under land conditions, because of its adaptations to tolerate the biotic and abiotic stresses prevailing under rainfed conditions. It is cultivated in arid and semi-arid areas of India. Crop diversification through arid fruits like ber ensured some production during the drought year when traditional crop fail. Crop production on dryland in general and marginal rainfed lands in particular results in low and unstable and often times uneconomic yields because of aberrant monsoon behaviour. Poor management of marginal lands cause the land degradation. The area under rainfed cultivation are facing some kind of land degradation or the other. These marginal lands are not able to sustain aerable crops during drought situation. Therefore, it is necessary to develop some alternate land use systems for theses lands. Tree component in dryland agriculture increases production and income besides imparting stability to the farming system. Fruit trees apart from theses advantages also yield valuable by-products like fodder and fuel wood, through their annual pruning and fruits, which are supposed to improve and maintain good health of human being.

In India, agroforestry has been a way of life. Our social, religious and cultural ethics have been closely linked with the planting and protection of various trees. Tree biodiversity has been preserved in many areas because of its role in farming and other land use systems. Integration of woody perennials in the farming systems is essential to improve land productivity through conservation, restoration of ecological balance, alleviation of poverty and mitigate risks of weather vagaries.

Table 3.10: Fruit Tree Species, Spacing and Number of Plants

Spacing (m)	*Tree Species*	*No. of Trees*	*Area (Acre)*
8 x 8	Aonla	40	0.64
8 x 5	Guava	250	2.50
6 x 6	Ber	40	0.36
6 x 6	Pomegranate	40	0.36
7 x 5	Citrus (Lemon)	125	1.10

Source: Dwivedi *et al.*, 2007.

Table 3.11: Returns from Fruit Based Agroforestry System on Farmer's Field

Species	*No. of Trees ha^{-1}*	*Fruit Yield (kg ha^{-1})*	*Returns from Fruits (Rs ha^{-1})*	*Crop Yields (kg ha^{-1})*		*Returns from Crop (Rs ha^{-1})*	*Total Returns (Rs ha^{-1})*
				Rabi	*Kharif*		
Aonla	86	6450	64500.00	3005	902	20507	95007.00
Ber	174	3915	15660.00	2980	975	31219	46879.00
Guava	125	5625	22500.00	2875	925	29931	52431.00
Citrus	57	2565	25650.00	3245	1015	33435	59085.00
Sole crop	–	–	–	3800	1278	40226	40225.00

Source: Dwivedi *et al.*, 2007.

Table 3.12: Returns from Various Components Under Agroforestry Systems

Species	*Return from Fruit (per cent)*	*Return from Crop (per cent)*	*Reduction in Crop Return (per cent)*	*Total Return (Rs ha^{-1})*
Aonla	67.9	32.1	24.2	95007.00 (136.2)
Ber	37.4	66.6	22.4	46897.00 (16.5)
Guava	42.9	57.1	25.6	52431.00 (30.3)
Citrus	43.4	56.6	16.9	59085.00 (46.9)
Sole crop	–	100	–	40226.00

Source: Dwivedi *et al.*, 2007.

Figures in parenthesis indicate per cent increase in total returns under agri-horti system over sole crop.

Table 3.13: Farmer's Opinion on Fruit Based Agri-Horti Land Use System

Particular	*Aonla (Emblica officinalis)*	*Ber (Ziziphus mauritiana)*	*Guava (Psidium guajava)*	*Citrus (Citrus aurantifolia)*
Bird damage	–	–	–	–
Theft	–	–	–	–
Crop rampage due to fruit harvesting	–	–	–	–
Watch and ward	Medium	High	High	Low
Shade effect on crops	Local	Local	Local	Local
Marketing demand in local market	Very high	Low	High	High
Time of fruit harvesting	November	February	January	January

Source: Dwivedi *et al.*, 2007.

In-situ Moisture Conservation

The survival of fruit seedlings can be improved by shaping the surface in immediate vicinity of fruit tree to collect the runoff into the root zone.

1. On levelled flat lands, 'Saucer' shaped depressions of 15 cm depth at centre and 1.0 m diameter are to be made.
2. On mild sloppy lands (2-4 per cent) 'V' shaped depression with 15 cm depth and 1.0 m diameter are to be made. The seedlings has to be planted near the apex of 'V'.
3. On sloppy lands (4-6 per cent) series of ridges are to be constructed with varied shapes (horse shoe or half moon or fish bone) on gentle gradient to interpret and store rain water.
4. Mulching with small stone pebbles or paddy husk on the surface of the basin with 5-10 cm thickness is useful to reduce moisture losses and better weed control.

In-situ water harvesting study was carried out to optimise catchment area per tree in arid zone (Anantpur- Andhra Pradesh) for maximum utilization of water by diverting the flow part artificially and naturally. The catchment area of 6 m x 15 m with 5 per cent gradient was recorded the highest fruit yield during 1997, 1998 and 1999 (22, 54 and 30 kg plant^{-1} respectively). It also enhance plant growth and quality of ber fruit *cv.* Gola (Reddy and Umamaheswari, 2003).

Water Management

There is need to protect the fruit seedlings during summer for better survival at least in the first year through protective irrigation. The limited water collected from runoff in ponds or in other sources can be effectively utilized through pitcher and portable drip system.

Pitcher System of Irrigation

Earthen pots of 8-10 lit. capacity having a hole made into bottom of the pot is to be placed at 30 cm depth and 15 cm away from the seedling in the basin at the time of planting the seedling. After filling with water, cover the pot with lid. Water drops near to the root through the vick, which will be useful for the growth of the seedling. The refilling of water in the pot can be done at interval of 7-10 days in summer.

Management Practices

Two weeding in a year are essential to improve the survival and growth of fruit seedlings. Stalking with sticks in rainy season is desirable to avoid wind damage and also for straight growth of seedlings. Tillage on subsequent summer monsoon shower facilitates better moisture conservation and weed control. Application of manure and fertilizer as recommended are to be followed regularly for maintaining fruit trees in healthy condition.

Recommendations (More, 2002)

1. To obtain the higher production and maximum profit in the soil upto 15 cm depth in the Scarcity Zone of Dhule in the rainfall zone 4, ber style or ber – bajra cropping system is recommended (MPKV, Rahuri).
2. In the light soils of Dhule region under Scarcity Zone, it is recommended to plant ber at 5.0 m x 2.0 m distance and grow bajra + pigeaon pea (2:1) in between two rows of ber (MPKV, Rahuri).
3. Pruning the four secondries between 1st April to 1st May is recommended for ber *cv.* Umran. Keep 60 cm branch and prun (MPKV, Rahuri).
4. Maximum yield can be obtained by pruning the ber trees in 15th meteorological week under Parbhani conditions. Pruning period can be prolonged from 15th to 17th meteorological weeks (MAU, Parbhani).
5. For higher yield of good quality fruits. April pruning of ber should be done in 2nd week of April (MAU, Parbhani).
6. For higher production of ber *cvs.*, Umran, Kadaka and Mukta, it is recommended to do pruning from 2nd to 4th week of April (i.e. in 15-17 meteorological weeks) (MPKV, Rahuri).
7. Mukta is a selection from local ber seedling trees, recommended in 1992. The fruits are attractive in colour and appearance. Average fruit weight is 7-10 g. Pulp is acidic-sweet is taste. Average yield of a mother tree is 150 kg and five years old budded tree gives 15-20 kg fruits. It is resistant to powdery mildew and fruit fly (MAU, Parbhani).
8. While planting of ber *cv.* Umran in the light soils of scarcity zone of Solapur, it is recommended to plant ber cv. Chhuhara to act as pollinizer along with *cv.* Umran (MPKV, Rahuri).
9. It is recommended to prepare the high value processed products such as sharbat and garvadi from ber. (PDKV, Akola)

Pests and Diseases

Insect Pests

There are six major insect pests described below:

- ☆ The greatest enemies of jujube are fruit flies. The flies prefer the largest sweet fruits.
- ☆ The larvae pupate in the soil and it has been found that treatment of the ground beneath the tree helps to reduce the problem. Control is possible with regular and effective spraying of insecticide.
- ☆ A leaf eating caterpillar *Porthmologa paraclina* and the green slug caterpillar *Thosea* spp. attacks the foliage.
- ☆ Mites, Larvacarus transitans forms scale like galls on twings retarding growth and reducing the fruit crop.
- ☆ Fruit fly is the most serious pest of ber. Infestation of fruit fly starts in September with the beginning of fruit set.
- ☆ Bark eating caterpillar- Gallaries of frassy web near the fork of branches indicated.
- ☆ Lesser pests include a small caterpillar, Meridarches scyrodes, that bores in to the fruit, the gray hair caterpillar, Thiacidas postica also Tarucus theophrastus, Myllocerus trasmarinus and Xanthochelus supecillous.

Diseases

The tree is subject to shrouding by a parasitic vine (*Cuscuta* spp). Powdery mildew (*Oidium* spp.) causes defoliation and fruit drop. Sooty mold (*Cladosporium zizyphi*) causes leaves to fall. Leaf spot results from infestation by *Cercospora* spp. and *Isariopsis indica*. In 1973, a witches broom disease caused by a Mycoplasma like organism was found in jujube plants near Poona University. It proved to be transmitted by grafting or budding diseased scions into healing *Z. mauritiana* seedlings. Leaf rust caused by *Phankospora zizyphivulgaris* range from mild to severe on all commercial cultivars in Punjab.

Fruits on the trees are attacked by *Alternaria chartarum, Aspergillus hasus A.parasiticus, Heliminthosporium atroolivaceum, Phoma hessarensis* and *Stemphyliomma valpara* disiacum. Twing and branches may be affected by *Enypella zizyphi, Hypoxylon hypomilltum* and *Patellaria atrata*. In storage, the fruits may spotted by the fungi, *Alternaria brassicicola, Phoma* spp., *Curvularia lunata, Cladosporium herbarum*. Rots are caused by *Fusarium* spp., *Nigrospora oryzae, Epicoccum nigrum* and *Glomerella cingulate*.

- ☆ Powdery mildew cause defoliation and fruit drop.
- ☆ Sooty mould, brown rot and leaf spot : Leaf spot results from infestation by *Cercospora* spp. and *Lsariopsis indica*.
- ☆ Witches-broom disease caused by a Mycoplasma like organism.
- ☆ Leaf rust caused by *Phakopsora zizyphi-vulgaris*.
- ☆ In storage, the fruits may be spotted by the fungi.

Table 3.14: Insect Pest, its Nature of Damage and Control Measures

Sl.No.	*Pest*	*Nature of Damage*	*Control Measures*
1.	Bark eating caterpillar (*Inderbela quadrinotata*)	Eats away the bark of the tree and cause sever damage which reduces the yield.	☆ The exposed holes should be painted with 10 ml monochrotophos 40EC or 40 ml of trichlorphon 50 EC. ☆ Application of a mixture of one litre kerosene and 100 g soap in 9 liters of water.
2.	Hairy caterpillar (*Euproctis froterna, Dasychira mendosa, Thiacidas postica*)	☆ The caterpillar feeds on leaves in initial stage ☆ The older caterpillars also feeds on the fruits and tender shoots and damage the trees. ☆ Start eating new foliage as it grows after pruning and this is continued by overlapping generations.	☆ In early stage, caterpillar can be controlled with 10 per cent HCCH powder. ☆ Or Trichlorphon ☆ Effctive control of 3rd to 6th instar larvae by 0.005 per cent Cypermethrin, 0.0014 per cent Delta methrin, 0.0075 per cent fluvalinate and 0.0005 per cent fenvalerate (Bhatnagar and Lakra, 1992).
3.	Leaf Chafer (*Adoretus pallens*), Lent bettle (*Adoretus decanus, Adoretus kanarensis, Adoretus stoliezkae*)	☆ Feeds on leaves during night. ☆ Pest is more active in the monsoon when new growth is produced. ☆ Leaves become like sieves. ☆ In severe cases, the whole tree is rendered leafless. ☆ Larvae hatch out in one week and feed on roots and vegetation.	☆ The bettles are controlled by spraying malathion 300 ml in 500 liters of water. ☆ Spraying of 0.2 per cent Carbaryl 50WP. ☆ The beetles can be trapped by using any source of light and killed by dropping them in to water containing Kerosene.
4.	Lac insects (*Laccifer lacca, Kerria lacca, Kerria sindica*)	☆ The shoots are affected with these small insects. ☆ They suck the sap from the branches which ultimately dry up. ☆ The small insects become active in summer. They secreate a thick, resinous substances which envelops their bodies. ☆ The secretion form a hard crust on the twings, which is collected to form a commercial resine. ☆ An infestation of 5000 nymphs/100 cm twings.	☆ Infested branches are cut off and destroyed. ☆ Tree should sprayed with Rogar 1 ml per liter of water or ☆ Demecron 1 ml in 3 litre of water. ☆ After pruning the trees should be sprayed with 0.1 per cent Dimethoate.

Contd...

Table 3.14–*Contd...*

Sl.No.	*Pest*	*Nature of Damage*	*Control Measures*
5.	Fruit fly (*Carpomyia vesuviana, Dacus correctus, Dacus dorsalis*)	✩ Infestation starts with onset of fruit setting. ✩ Adult female lays eggs singly by inserting its ovipositor in the young developing fruit. ✩ After 2-5 days, the larvae hatch start feeding on the pulp and make galleries. ✩ Accumulated excreta of the larvae in galleries sometimes result in rotting of the fruit. ✩ Fruit drops.	✩ Cultivation of orchard soil and destroy the hibernating pupae by exposing them to bright sunlight and birds (Singh *et al.,* 1973a, Chundawat and Srivastav, 1978 and Bakshi and Singh, 1974). ✩ Three application of or two sprays of malathion (Chundawat and Sriwastav, 1978 and Patel *et al.,* 1989). ✩ 0.005 per cent Fenvalerate or 0.0015 per cent Deltametrin or Phosphamedon (Bagle, 1992). ✩ Two applications with 0.1 per cent Dichlorophos, one at pea stage of fruit and another after one month (Raghumoorthy and Arumugam, 1992). ✩ Biological control *Bracon fletcheri, Opius carpomyiae* and *Omphalina* spp. have been recoded two parasitise fruit fly.
6.	Fruit borer (*Meridarchis scyrodes*)	✩ Mainly in southern and weastern India the redish larvae bore into the fruit and feed on the pulp (Sonavane and Dorge, 1971, Pareek and Nath, 1996).	✩ First spray at pea stage with Monochrotophos (0.03 per cent), Second spray after 15 days and Third spray 15 days after with 0.01 per cent Carbaryl has been recommended (Pareek and Nath, 1996), (Singh *et al.,* 1973 a), (Chundawat and Srivastava, 1978) (Bakhshi and Singh,1974), (Chand *et al.,* 1986).
7.	Mites	✩ Produce galls in floral buds preventing fruit production (Yamdagni and Gill, 1968). ✩ The mites (*Eriophes cernus*) occures in India through out the year (Mukherjee *et al.,* 1994). ✩ Galled tissues contain higher total carbohydrates and reducing sugars and more α amylase activity as compared to the normal tissue (Tandon and Arya, 1979).	✩ Treatment with 0.04 per cent Dicofol gave the best control.

Table 3.15: Diseases, its Nature of Damage and Control Measures

Sl.No.	*Disease*	*Nature of Damage*	*Control Measures*
1.	Powdery mildew (*Oidium* spp., *Odium crysiphodes*)	☆ Initially white spots of fungus appear on the leaves, fruits etc. In case of severe infestation the powdery mass spreads all over the surface of leaves and fruits. ☆ The fruit drop prematurely or becomes sunken, misshapen and corky. ☆ Causing a marked reduction in yield. ☆ Symptoms appears on flowers and newly set fruits. ☆ White powdery spots on the surface of the fruits and later covers the whole fruit surface. ☆ The spots turn into light brown to dark brown discolouration. ☆ Infected area becomes slightly raised and rough.	☆ Use of resistant cultivars. ☆ Spraying of Karthane (dinocap) 0.2 per cent or morestan (quinomethionate) at 0.05 per cent for the control of powdery mildew. ☆ In addition to Karathane, bavistin at 0.05 per cent or wettable sulphur (0.2 per cent) at 10 days interval in October and December. ☆ Dusting with sulphur @ 150 to 200g per tree and subsequent 3 dusting at an interval of 15-20 days. ☆ Spraying of Dinocap (0.2 per cent Karthane WP or 0.1 per cent Karthane EC), 0.2 per cent Sulfex (Singh and Sindhu, 1985) or Tridcomorph (Singh *et al.*, 1995).
2.	Sooty mould or black spot (*Isariopsis* spp., Gupta and Madaan, 1977b)	☆ Appearance of black spots on the lower surface of the leaves and in advanced stages the entire lower surface may be curved by sooty mould and the leaves may drop down. ☆ Characterised by dark gray spots on lower surface of the leaves, affected leaves may fall off. ☆ Characterised by sooty tuft like circular to irregular black spots on the outerside of the leaves, it covers the entire lower surface giving a sooty appearance. ☆ The leaves show yellowish and brownish discolouration on the upper surface and drop prematurely.	☆ Spraying the trees with Dithane Z-78 (Zineb) 0.3 per cent or 0.6 per cent copper oxychloride (blitox), or 0.05 per cent defoliation at fortnightly interval. ☆ Spraying Dithane M-45 (0.25 per cent) or Foltart (0.1 per cent) as soon as the disease appear. Subsequent 2 to 3 sprays be given at an interval of 15 to 20 days depending upon intensity of disease. ☆ Disease can be controlled by Captatol (0.2 per cent), Carben dazin (0.1 per cent), Mancozeb (0.2 per cent) and Copper oxychloride (0.2 per cent). The spray should be applied 2-3 times at 15 days intervals starting with the first sign of symptoms (Pareek and Nath, 1996). ☆ 0.1 per cent Carbendazin and 0.2 per cent chlorothaltrill (Rawal and Saxena, 1989). ☆ 0.3 per cent Zineb or 0.6 per cent Blitox (Sharma *et al.*, 1983).

Contd...

Table 3.15–*Contd...*

Sl.No.	*Disease*	*Nature of Damage*	*Control Measures*
3.	Fruit rot (*Alternaria alternata*)	☆ Light gray brown spots appear on the lower end of the fruit. ☆ Some times dark brown rings are formed on fruits. ☆ Slightly depressed, brown to dark brown, circular lesions appear on the fruit. Sometimes concentric rings are also present on these spots.	☆ Spraying Dithane Z-78 (Zineb) or blitox 0.2 per cent at the early stage of infection (Chand *et al.*, 1986).
4.	Phoma fruit rot	☆ Appears on ripening fruit. ☆ The infected fruits remain small and developed slightly depressed, dark brown spots near the stem ends. ☆ The lesions become irregular in shape and measure 15-25 mm in diameter.	☆ The disease can be controlled by spraying of 0.2 per cent Dithane Z-78 or Dthane M-75 (Chand *et al.*, 1986).
5.	Collectotrichum fruit rot	☆ Appears at the start of ripening of the fruit in the form of small, slightly depressed light brown, water soaked lesions. ☆ These spots coalesce and enlarge.	☆ The disease can be controlled by 2-3 sprays of 0.2 per cent copper oxychloride at 3 week intervals (Chand *et al.*, 1986).
6.	Trichothecium fruit rot	☆ The disease is observed during the spraying in the form of pink spots on the fruits. The fungus can survive in the soil for a long time.	
7.	Cladosporium fruit rot	☆ The disease appears near the time of fruit ripening. ☆ Injured fruits become infected. The symptoms of the disease start from the tip of the fruit forming light brown to dark brown spots. Later, a greenish fungal growth is also seen on these spots.	
8.	Rust (*Phalcospora ziziphi*-vulgaris) (Gupta *et al.*, 1984)	☆ The disease appears towards the end of winter on the leaves in the form of small, irregular, reddish brown uredopustules on the lower surface. ☆ Pustules aggregate on the tips and margins resulting in large necrotic spots and death of tissues. ☆ The infection advances over the white surface and the infected leaves become dry and fall off.	☆ The disease can be controlled by application of 0.4 per cent copper oxychloride or 0.2 per cent Zineb or Captafol (Pareek and Nath, 1996). ☆ In Chinese jujube, the disease severity was reduced by 84.9–77.5 per cent after application of metalaxyl and Bordeaux mixture (Sun *et al.*, 1989).

Contd...

Table 3.15–*Contd...*

Sl.No.	*Disease*	*Nature of Damage*	*Control Measures*
9.	Witche's broom	✰ Witche's broom affects both Indian and Chinese jujubes (Kim, 1965, Zhu *et al.*, 1983). ✰ Pandey *et al.* (1976) first reported the disease in ber. The affected trees become weak and produce small, yellow leaves typical of other phytoplasma disease. ✰ The disease is transmitted to healthy seedlings by budding and grafting with diseased scions.	✰ The disease can be controlled by treatment with 500 and 1000 ppm oxytetracycline. ✰ La *et al.* (1977) recommended control through transfer of 1000 ppm tetracycline hydrochloride into the trees from plastic reservoir through plastic tubes. Wang *et al.* (1980) successfully controlled it by two commercial formulations of oxytetracycline.

- ☆ Fruit rots are caused by *Fusarium* spp., *Nigrospora oryzae, Epicoccum nigrum* and *Glomerella cingulate.*
- ☆ Alternaria leafspot is characterised by dry spots on lower surface of the leaves.
- ☆ Fruit rot- light gray-brown spots appear on the lower end of the fruit.

Recommendations (More, 2002)

1. Wettable sulphur (0.2 per cent) is recommended to control of powdery mildew disease of ber. Give the first spray after the pruning and the second one after a month. Subsequent sprays of fungicides should be given at an interval of 20 days starting from the initiation of disease. (MPKV, Rahuri).
2. For control of powdery mildew disease of ber dusting of sulphur or spraying of sulphur (0.2 per cent) is recommended. One prophylactic spray before flowering and subsequent seven sprays after the fruit set are recommended. (MPKV, Rahuri).
3. One the basis of results obtained at Sindewahi, it is recommended to spray Bayleton (0.15 per cent) or Calixin (0.15 per cent) or Karathane (0.15 per cent) in the 2nd week of September. Subsequent spraying of any of these fugicides should be repeated at an interval of 15 days till the maturity of fruits to control the powdery mildew. (PDKV, Akola).
4. For control of powdery mildew disease of ber, give five sprays of 0.2 per cent sulphur (WS). First spray should be given at 50 per cent flowering as preventive measure. Remaining four sprays should be given at an interval of 20 days as soon as the disease is appeared. (MPKV, Rahuri).
5. For control of powdery mildew disease of ber, give five sprays of 0.2 per cent sulphur or four times dusting of Sulphur (330 mesh) @ 20 kg ha^{-1} is recommended to control the powdery mildew of ber. First application should be made start at the initiation of disease on the flower or small fruit lets and subsequent sprays should be given at an interval of 15 days. (PDKV, Akola).
6. Three sprays with Fenvelrate at 0.01 per cent or 0.002 per cent Deltamethrin at 30 days interval starting from fruit set are advocated for the control of ber fruit borer. (MPKV, Rahuri).
7. A spray schedule comprising of three sprays at an interval of 30 days starting from fruit set either Fenvelrate 0.01 per cent, Deltamethrin 0.02 per cent, Carbaryl 50 WP 0.2 per cent and Quinolphos 0.05 per cent is found to be the most appropriate to control of ber fruit borer. (MPKV, Rahuri).
8. For control of the ber fruit borer, the insecticides *viz.,* 0.002 per cent Deltamethrin, 0.01 per cent Fenvelrate and 0.05 per cent Quinolphos are recommended. First spray should be given when the fruits are of peanut size and secondary spray 15 days after the first one. (MPKV, Rahuri).
9. To control the ber fruit borer, it is recommended to spray 0.02 per cent Carbaryl (50WP) or 0.025 Dipluveran or karanj extract (2 per cent) in

addition to the earlier recommended insecticides such as 0.01 Fenvelrate or 0.002 per cent Deltamethrin. Spray thrice after the fruit set at a interval of 30 days. (MPKV, Rahuri).

10. For the control of ber fruit fly, traps made by using Methyl eugenol should be used. A trap is having 3 ml Methyl eugenil is effective for 45 days. Four such traps per hectare control the fruit fly effectively. (BSKKV, Dapoli).
11. For control of the chafer beetles (leaf eating beetles) of ber the insecticidal spray of Carbaryl 50 WP at 0.20 per cent is advocated to control the chafer beetle. (MPKV, Rahuri).

Table 3.16: Maturity indices for Cultivars of Ber

Cultivar	*TSS (per cent)*	*TSS/Acid ratio*
Gola	16.7-17.0	76-84
Kaithli	16.2-16.5	75-100
Umran	17.1-17.4	93-100

Source: Bhatia and Gupta, 1985.

Ber Fruits

Buddiing Method in Ber

1. Budwood

2. Rootstock prepared for patch or shield budding

3. Scion buds tied after insertion

4. Scion buds successfully sprouted

Training in Ber

1. First year age ber plant **2. Second year age ber plant**

3. Three year age ber plant

Pruning in Ber

Moisture conservation in Ber

1. Mulching

2. Semicircular basin for *in situ* moisture conservation

4

Pomegranate

S.M. Todmal, S.D. Kale, A.B. Pawar and A.D. Kadlag

Department of Soil Science and Agril. Chemistry, MPKV, Rahuri, Maharashtra

---

Pomegranate (*Arnica granatum* L.) is an important fruit of semiarid and arid regions of India. Its cultivation is gaining importance because the plant is hardy and can tolerate drought and alkalinity. There are two types of pomegranate grown in India. The edible varieties can be broadly classified into two groups namely soft seeded or so called seedless and hard seeded.

The area under pomegranate in the country is increasing at a faster rate due to its hardy nature, low maintenance cost, low water requirement, high yield potential, good keeping quality and versatile adaptability.

Pomegranate is a fruit grown mainly in America, Azerbaijan, Iran, India, drier parts of Southwest Asia. Peninsular Malaysia, the east India and Tropical Africa, along with parts of California and Arizona. However, it has been the native fruit of the region from Afghanistan, Pakistan and Iran to the Himalayas in Northern India. Pomegranate has been associated with a high nutritional value and a large number of health benefits, illustrating the reason why it has been cultivated in the Mediterranean region as well the Caucasus since ancient times.

Screening of new cultivars introduced from other countries is being done with a view to testing their adaptability and other characteristics. The crop can be successfully introduced in areas such as Solapur and Ahmednagar districts where congenial climate and irrigation facilities exist during summer months.

Pomegranate is an important commercial fruit in Iran which is mainly grown in Yard Markazi, Khorosan and Fars provinces. Pomegranate requires a long hot summer for fruit to mature. However, it can withstand low temperatures in winter. It is a drought and salt tolerant plant. Pomegranate flowers from late April until mid

May. Whereas, maturity and ripening extends from mid September to November depending on the cultivars (Patil and Karale, 1990).

Pomegranate is high value crop, mostly used as fresh fruit. It has a long shelf life and is an ideal fruit for long distance transportation and prolonged storage. A part from its demand for fresh fruits and juice, wine and candy also gaining importance in world trade.

The pomegranate contains potassium, vitamins A, B (Niacin – B_3, pantothenic acid - B_5, folic acid - B_9), vitamin C, vitamin E and iron. The pomegranate antioxidant properties are the most well known. Some studies shown that the juice of the pomegranate has about three times the antioxidant power of green tea and red wine. Common antioxidants in the pomegranate are vitamin C and vitamin E other antioxidants contained in the pomegranate are :

- ☆ *Polyphenols*: Phyto-chemicals (plant chemical) or phytonutrients that are rich in antioxidants.
- ☆ *Anthrocyanin* : Polyphenols that give the pomegranate its deep red colour.
- ☆ *Tannins* : Polyphenols that give the pomegranate its colour and tart taste.
- ☆ *Punicalagin* : A hydrolyzable tannin.
- ☆ *Ellagic acid* : Punicalagin breaks down in to ellagic acid.

The nutrients in pomegranate reduce the risk of heart disease, help to prevent plaque build-up in your arteries by reducing the oxidation of LDL (bad cholesterol), reduce inflammation, be effective against osteoarthritis, help to prevent prostate cancer and other cancers, enhance immune function promote production of red blood cells, expel tape worms, strengthens the bladder, strengthen the gums, reduce diarrhea.

Being rich in minerals, tannic acid, flavonoids and proteins (Chavan *et al.*, 1995), rind of pomegranate fruit is dried and its powder used as medicine, for making tooth paste and in industries for leather tanning and dying (Salunkhe and Kadam, 1995). Pomegranate rind is a powerful astringent, it is administered in the form of decoction for diarrhea and dysentery, often combined with opium and as an injection in case of leucorrhoea. In addition to most of the important vitamins, proteins and flavonoids, the pomegranate rind powder is rich in minerals like Ca, Mg and Fe and are important minerals required for healthy living. The rind powder is rich in β-carotene. Due to the protective effect as an antioxidant, β-carotene has come into more true light in new discoveries that implies into more benefit and heavy commercial promotion. Under these circumstances, there is every need to produce more mineral and nutrient rich products for using as protectants, nutrition plays an important role in influencing the yield and quality of fruits. Yet due attention has not been paid to standardize the nutritional requirement of pomegranate.

Pomegranate is highly remunerative crop for replacing subsistence farming, alleviating poverty. There has been a steady increase in area and production of pomegranate in the country. The total production in India during year 2008-09 was 992000 MT on an area of 132500 hectares with productivity of 10.30 t ha^{-1}. In

Maharashtra pomegranate production during 2008-09 was 844551 MT on an area of 96406 hectares with productivity of 10 t ha^{-1}. Average world production is 10-12 t ha^{-1}. Export of fruit - 12034.54 MT, and valued of Rs. 25.87 crores.

By 2025 the area under pomegranate could increase to 7.5 lakh ha, from 1.25 lakh ha, at present. Consequently production is expected to increase by 10 fold and export by 6.97 fold by 2025. To achieve these targets co-ordinated and sustained efforts are required by all concerned with pomegranate research and development to develop scientific, profitable ecofriendly and innovative technologies for cultivation both in traditional and non traditional areas.

Soil, plant, water and fertilizers arc the major ingredients of interest in exploiting the higher yield with export quality. The fertilizers are invariably used for supplying the nutrients to growing plants. The use of inorganic fertilizers, organic manures and biofertilizers individually will not serve the purpose of higher yield with quality of crop. The very rational approach to achieve the quality produce with adequate yield of pomegranate is balanced STCRC approach of nutrient management.

The sustainable production is only possible by proper use of inputs for maintaining the soil fertility. The gravity of environmental degradation caused by faulty and intensive agricultural practices has become evident and there is increasing pressure to develop ecologically sound, viable and sustainable soil fertility management practices through integrated nutrient management (INM).

The basic concept of STCRC is the improvement in soil productivity on long term basis through approach use of fertilizers, biofertilizers and organic manures including green manures and their scientific management for optimum growth, yield and quality of crops and cropping systems in specific agro-ecological situation. INM helps to restore and sustain soil fertility and productivity. It may help to check the emerging deficiency of nutrients. Further, it brings economy and efficiency in fertilizer use and improvement in physical, chemical and biological environment of soils.

Integrated nutrient management strategies emphasized on reduction of nutrient losses from applied chemical fertilizers, retention of soil nutrients, alternate or supplementary source of nutrients. The supplementary sources of nutrients includes organic manures and biofertilizers plays important role. The organic manures *viz.*, FYM, compost, vermicompost. poultry manures, oil cakes (castor, karanj, neem cake etc.) are well known to improve physical chemical and biological properties of soil, ultimately crop yield. Crop residue management, green manuring, legumes as intercrop etc., are the key components of INM.

Practicing the integrated nutrient management (INM) is a holistic management approach that is ecologically, socially and economically viable and at the same lime sustaining productivity and ensuring high level of production. The scientific integration of mineral fertilizer with organic sources of plant nutrients alongwith biofertilizers are the only future viable options for sustaining soil productivity under the varied agro-ecological situations, where maintaining soil fertility with less costly options is an impressive.

Origin and Distribution

The pomegranate tree is native from Iran to the Himalyas in northen India and has been cultivated since ancient times throughout the Mediterranean region of Asia, Africa and Europe. The fruit was used in many ways as it is today and was featured in Egyptian mythology and art, praised in the old testament of the Bible and in the Babylonian Talmud, and it was carried by desert caravans for sake of it's thirstquenching juice. It traveled to central and Southern India from Iran about the first century A.D. and was reported growing in Indonesia in 1416. It has been widely cultivated throughout India and drier parts of southeast Asia, Malaya, the East Indies and tropical Africa. The most growing regions are Egypat, China, Afganistan, Pakistan, Bangladesh, Iran, Iraq, India, Burma and Saudi Arebia. There are some commercial orchards in Israel on costal plain and the Jorden Valley.

It is rather commonly planted and has become naturalized in Bermuda where it was first recorded in 1921, but only occasionally seen in the Bahamas. It is grown for it's fruit mostly in the dry zones. The pomegranate continued to be dispersed around the globe, reaching china by 100 BCE (Anarinco, 2006) By 800 CE, the fruit was spread throughout the Roman empire including Spain. At the same time, it was known to be extensively cultivated in Central and southern India (Morton, 1987). By the early 1400's the pomegranate had made its way to Indonesia (Morton, 1987). In the 1500's and 1600's the Spanish introduced this species to central America, Mexico and South America (LaRue,1980). The first clear evidence that the pomegranate was in the area to become the United States was in the early 1700's when they were grown in Spanish, Florida and English Georgia. By 1770, the pomegranate made its way to the West coast and was growing in the mission of California (Morton, 1987).

Climate

Pomegranate has a versatile adaptability to wide range of climatic conditions. It can grow in plains as well as on hills up to an elevation of about 1850 meters. This fruit tree grows the best in semi arid climate where cool winter and hot summer prevail. In areas of low temperature, the tree is deciduous but in tropical and subtropical conditions it is evergreen or partially deciduous. The tree requires hot and dry climate during the period of fruit development and ripening. The optimum temperature for fruit development is about 38 °C. The tree can not produce sweet fruits unless the temperature is high for a sufficiently long period. The quality of fruit is adversely affected in humid climate.

Pomegrate is winter hardy and drought tolerant plant and can thrive under desert condition. Although it is highly drought resistant, the plant bears well only under irrigation. Two year old shoots of hardy cultivars were not damaged even at 2°C. Aridity and frequent anomalies of the climate cause leaf shedding and fruit cracking.

Soil

The pomegranate is not very particular about its soil rquirement and can be grown on diverse type of soils. The tree gives very good yield in deep loamy or alluvial soil, although it thrives well on comparatively poor soils where other fruits fail to flourish. It can tolerate soil with which are slightly alkaline, though it can also be grown in medium or light black soils of at least 60 cm depth. It is salt hardy fruit trees.

Sandy loam to deep loamy soils with good drainage are ideal for cultivation. It performs well on light soils and affaince good fruit colour and quality. It is tolerant to salinity and can be grown on slightly alkaline soils. Patil and Patil (1982) observed 20 per cent growth suppression at 25.5 ESP and this seems to be limit up to which pomegranate can be grown. Salinity reduces the number of roots per plant. The pomegranate thrives on calcareous alkaline soil and on deep acidic loam and a wide range of soils in between these extremes. In northern India, it is spontaneous on rockstrewn gravel.

Propagation

Propagation by seed is discouraged because of the plants raised from seeds do not bear fruit of uniform quality and yield low and poor quality fruits. The commercial vegetative method of propogation is by cutting (hard wood cutting). The cutting, preferably 8 to 10 cm long, should be obtained from fully mature wood about one year old. The best time of making the cutting is December-January when the plants shed leaves. The cuttins are planted in the nursery fields with not more than one third of the cutting being exposed. The plants are usually realy for transplanting within nine months. Ground layering and root suckers are other methods used for multiplying the pomegranate plants.

The hard wood cuttings treated with 50 $\mu g\ g^{-1}$ (ppm) indole butyric acid (IBA) and planting at moisture level of 15.95 per cent greatly enhances root development and survival. The cuttings are set in beds with 1 or 2 buds above soil for 1 year and then transplanted to the field. Grafting has never been successful but branches may be air layered and suckers from a parent plant can be taken up and transplanted.

Many pomegranate plantations in India have been raised from seedlings but the trees from seedlings show wide variation in characters. Low yields and poor quality of fruits are mainly due to seed propogation. In order to obtain fruits of uniformly good quality, it is essential to establish orchards with vegetatively propogated plant. The common method of vegetative propagation is by rooting hardwood cuttings. The cuttings are taken from fully mature wood, about one year old and suckers which arise from the base of the stem are usually utilized for this purpose, 20-25 cm long cuttings showed good rooting (Rajanana, 1981). Root formation is markedly improved by using growth substances. Phadnis (1974) reported higher percentage of rooting and large number of roots by treatment with IAA at 200 ppm. Basal and medium cuttings rooted to the extent of 75 per cent by soaking the base of cuttings in 50 ppm IBA for 18 hours and to 66 per cent by quick dip of 20 seconds in 5000 ppm IBA (Sulladamath, 1985). Bose and Mandal (1972) obtained 85-100 per cent rooting in pomegranate using IBA at 3000 ppm in talc under mist and the rooted cuttings also showed very high percentage of survival. Softwood cuttings of pomegranate when treated with 0.5 per cent potassium salt of IBA as quickdip and planted at fertile media under intermitant mist gave high percentage of rooting.

Moisture level in the rootings media improves the rooting percentage to significant level. The rooting percentage increased from 39.2 to 90.0 per cent by increasing moisture level of the media from 5.80 to 20.14 per cent.

Pomegrante may also be propogated by air layering or *gootee*. Treatment with 1000 ppm IBA in lanolin as carrier was found to improve rootings. Anther culture of pomegranate was reported by Moriguchi *et al.* (1987). Anthers when cultured on nutrient medium containing BA and NAA showed formation of callus from the anther walls after 30 days. Adventitious shoots were produced by transferring the callus to a medium containing lower concentration of BA and NAA.

Species and Cultivars

The genus has two species i) *Punica protopunica* ii) *Punica granatum. Punica protopunica* is found wild in the Scotoria island and *Punica granatum* is cultivated in tropical and subtropical parts.

Punica granatum has been classified into two subspecies, chlorocapra and porphyrocarpa. Chlorocarpa is found in Transcaucacus region and porphyrocarpa in central Asia.

Cultivars

The cultivars, Schalvar which has thin rind and robust with thick rind are grown in India (Firouzeh and Ladders, 1979). Of the 46 culivars evaluated for the gravelly soils of the Fergana Valley in Uzebekistan the promising ones were Bala-Myursal. Several types of pomegranates cultivated in India are distinguished by shape of the fruit, colour and thickness of rind, colour of the seeds (aril) and taste.

On the basis of yield and physicochemical characters, various cultivars have been recommended for the commercial cultivation in different states in India *viz.*, Ganesh and Muskat in Maharashtra. Dhoka in Gujarat, Jellore seedless, Jodhapuri Red, Jodhapuri white and seedless (Bedana) in Rajasthan, Bassein seedless and Madhurigi in Karnataka, Chawla, Nabha, Country Large Red in Haryana (Hissar).

In northen India, there are no well recognized commercial varieties of pomegranate. In western India two types, locally known as Alandi and Dholka are grown. The different cultivars of pomegranate grown in India are shown in Table 4.1

Pomegranate Cultivars Commonly Grown in State and Particular in Solapur

The major varieties cultivated in Maharashtra are GBG-1 (selection from Dholka). GBG-1 was later on named as Ganesh. Further, selected elite plants from Ahmednagar naming it as Muskat variety. The Jyoti was released from Banglore. The blood red colour of the Russian types was incorporated in Ganesh and a promising line from F_2 population (No. 61) combining all desirable quality attributes has been released with named Mridula (Ganesh X Gulsha rose pink). The Bhagwa was developed by MPKV, Rahuri. In Maharshtra the popular varieties are Bhagwa, Ganesh, G-137, Arakta, Mridula, Dholka, Alandi and Pune Araka. The cultivars of pomegranate growned in solopur districts are Bhagawa, Ganesh, G-137, Mridula

Table 4.1: Primary Characteristics for Pomegranate Cultivars Indicated as "Important" in the Literature

Cultivar	*Fruit Size (g)*	*Traits*	*Origin*	*Reference*
Agriducle or Ojobs 4	524	Red arils with hard seeds, bitter/sweet, medium acid	Spain	Amoros *et al.,* 2000
Alandi (or Vadki)		Deep pink arils and very hard seeds, sweet/sour	India	Morton, 1987
Asinar	505	Large fruit, red arils sweet/sour soft seeds	Turkey	Gozlekci and Kuynak, 1997
Bedana		Medium-large, brownish white rind, pink-white arils, seeds soft, sweet	India	Morton, 1987
Borde or Albatera	370	Deep-red arils with hard seeds, bitter, high acid	Spain	Amoros *et al.,* 2000
Dholka		Large yellow-red rind, white arils and hard seeds, sweet	India	Morton, 1987
Early foot hill		Deep red arils, medium hard seed sweet/sour	USA, 2-4 weeks earlier than wonderful	LaRue, 1980
Early Wonderful		Deep red arils medium hard seeds, sweet/sour	USA, 2-4 weeks earlier than wonderful	California Rare fruit growers, 1997
Ekasilik		Sour (5 per cent TA), red arils	Turkey	Gozdekci and Kayanak, 1997
Emar		Dark red skin, red arils, sweet with low TA	Turkey	Gozdekci and Kayanak, 1997
Eversweet		Pink to red fruit with pink arils, soft, seeds sweet even immature	USA	Dave wilson Nursery, 2005, Karp, 2006
Ganesh		yellow-pink rind and pink-red arils, Very soft weeds, sweet/sour	India	
Golden Globe	Very large	Golden green fruit with pink bluish, pink to red arils, soft small sweet	USA	Karp, 2006
Garanda		Deep red arils, medium hard seeds, sweet/sour	USA, redder, 1 month earlier sport of 'Wonderful'	California Rare Fruit Growers, 1997
Hicaznar		Dark red skin, red arils, sweet/sour	Turdey	Gozlekci and Kaynalk, 1997
Kandhari (also called Arakta)		Large fruit, deep-red rind, with deep pink to blood-red arils, hard seeds, sweet/sour	India	Morton, 1987

Contd...

Table 4.1–*Contd...*

Cultivar	*Fruit Size (g)*	*Traits*	*Origin*	*Reference*
Katirbasi	517	Large fruit, large red arils, sweet/sour	Turkey	GozlekciandKayanak,1997
Mollar de Eleche 15	272	Deep red arils with soft seeds, sweet, low acid	Spain	Amoros *et al.*, 2000
Mollar de Orihuels	414	Red-pink arils with soft seed sweet, low acid	Spain	Amoros *et al.*, 2000
Pinon Tiemode Ojos 9	405	Red-pink arils with soft weeds, sweet low, acid	Spain	Amonos *et al.*, 2000
Valenciana		Small, early, but not top quality	Spain	Costa and Melga rejo, 2000
Wonderful		Deep red arils, medium hard seeds (we would call medium-soft). Sweet/sour	USA	Morton, 1987

and Phule Arakta. More than 500 cultivars of pomegranate have been named (IPGRI, 2001) but such ancient and widespread fruits often have considerable synonymy, in which the same basic genotypes is known by different names in different regions. Synonymy is likely further encouraged by the fact that husk and aril colour can vary markedly when grown in different regions. A number of characteristics vary between pomegranate genotypes and are key to identification, consumer preference, preferred used and potentially niche marketing. The most important traits are fruit size, husk colour (ranging from yellow to purpule, with pink and red most common), aril colour (ranging from white to red), hardness of the seed, maturity, juice content, acidity, sweetness and astringency.

Because of differences in quality and productivity observed in commercial planting. Spanish researchers have selected district clons of their most important cultivars. Selections were made in 1986 in the provinces of Alicante and Murica. Numberd clones were propogated and are undergoing replicated trials on identify the best materials (Amoros *et al.*, 2000).

The countries of pomegranate origin grown innumerable cultivars with many regional favorites. Local pomegranate germplasm collections have been established in several Mediterranean countries including Spain, Morocco, Tunisia, Greece, Turkey and Egypt (Mars, 2000). India has three collections containing at least 30 accessions each (Gulick and Vansloten, 1984). Azerbaijan, the Ukraine, Uzbekistan and Tajikistan have collections of 200 to 300 accessions and the collection of the Turkmenistan Experimental Station of Plant Genetic Resources is the largest in the world, containing more than 1000 accessions (Levin, 1995).

Little detailed information is available on pomegranate grown outside of western Europe and the United States, although efforts have been made to assemble cultivar summery information for this paper (Table 4.1). Cultivars mentioned as important in the literature but with no description, include 'Anmar', Aswad and Halwa from Iraq, Mangulati, from Saudi Arabia, and Red Loufani and Rasel Bahl from Israel and Polestine (Morton, 1987).

Pomegrante cuttings root with greate ease, facilitating spread of desirable clones. Efforts to graft pomegranate are reported not to be successful, but air-layering and root-sucker transplantation can be used for vegetative propagation (Morton, 1987).

Planting and Spacing

The planting of the rooted cuttings in the main field may be done during monsoon. The plants are removed from the nursery with a ball of earth and most of the leaves as well as the side shoots are pruned off. The layout is done according to square or hexagonal system and planting is done 3 to 6 m. apart. Light irrigation should be given immediately after planting. Rooted cuttings or air-layers are planted in pits at the required distance and filled with 20 kg farmyard manure and fine soil.

The layout is done following square or hexagonal system. Pits aoubt 60 cm^2 are dug at the required distance and filled with mixture of organic manure and soil. The planting material are planted in these pits during monsoon. Generally, in the poor soils of Deccan, the plants are given a spacing of 3.5 – 5.0 meters; in lighter but deeper soils the distance is increased up to 5x5 meters.

Planting density is the most important yield contributing factor which can be manipulated to attain the maximum production per unit area. The optimum spacing is important for the maximum utilisation of land and good income over a long period. Faiziev (1973) conducted aspacing trial in pomegranate with cultivars Kazakai, Ak Dona and Achik Dona using 5x3, 4 x 2.5 or 4 x 1.5 m^2 spacing which gave 666, 1000 and 1666 plants ha^{-1} respectively. The highest yield was obtained from closets spacing. Planting density had no adverse effects on fruit weight. Kutuzova, (1984) recommended spacing of 3 x 2.5 or 4 x 2 m^2 (1333 and 1250 plants ha^{-1} respectively) for high and economic yield of good quality fruits.

In an investigation at the Mahatma Phule Agricultural University (MPAU), Rahuri (India) with a plant population of 1000, 666, 500 and 400 plants ha^{-1} respectively, observed that as the plant density was increased, yield per hectare also increased without affecting fruit quality, and a density of 1000 plants ha^{-1} gave 2.30 times higher yield and 2.44 times more profit compared with normal plant population of 400 plants ha^{-1}. It was also recommended that for higher yields for the first four to five years after planting a distance of 5x2 m^2 may be adopted and alternate plants may be removed afterwards maintaining a planting distance of 5 x 4 m^2.

Normal spacing adopted is 5x5 m^2 on marginal and very light soils. It can be reduced to 4.5 m. Close spacing (5.0 x 2.5 m^2) results in higher yields, but hazardous in later periods (Bagate, 1979) and requires removal of alteranate plant after 6-7 years to maintain spacing of 5 x 5 m^2. Spacing of 3.0 x 2.5 m^2 or 4.0 x 2.0 m^2 produced high and economic yields of good quality (Kutuzova, 1984) in erect growing, deciduous varieties.

Training and Pruning

At planting, reduce each plant to a single stem and shorten this to about 50 cm above ground level. A short stake may be needed for support until the stem becomes rigid and sufficiently strong; Subsequently, select three to five shoots arising from the upper half of the stem to provide the main frameworks. These will be shortened at each winter pruning to produce a strong compact frame. Several secondary limbs shoud be allowed to develop from each main limb, but excessive numbers which would lead to over crowiding should be removed, as should any suckers which develop from the base of the tree.

The natural growth habit of the pomegranate is to produce many suckers from the base of the trees. If a single truck tree is desired, only one vigorous suckers or the trunk of the original nursery tree should be selected and branches grown from it. Basal suckers should be removed periodically to promote growth from the main trunk of the newly planted tree. If the orchard trees are to be develoed into a multiple trunks system, five or six vigorous suckers should be selected around the base of the young tree and allowed to grown. Selection of the five or six suckers to be developed into permanent trunks may take two or three or more years until good trunks are correctly positioned to form a sturdy and symmetrical tree. All other suckers should be removed in summer and during dormant pruning.

Many growers prefer the multi trunk system. Incase of frost injury, usually only one or two trunks are injured, leaving the others to continue bearing. New trunks can be trained from suckers and full production restored to the tree in 2 or 3 years. Single trunk trees may be completely killed except for suckers coming from the ground. Trees trained to a multiple trunks require less frequent case in pruning during the first few years and come into bearing sooner than trees having only one trunk.

Pomegrante plants allow to grow as bush with a number of main shoots arising at ground level and growers find it convenient to allow 4-5 such shoots to develop, though previously single stem training was advocated. To attain this, all the side shoots should be removed at the time of planting, the main stem is pinched at a height of about one meter and upto 25-30 cm below the cut surface, 4-5 branches well distributed on all sides should be allowed to grow to form main limbs of the tree. The desired shape of the tree is formed within 2 to 3 years of planting.

Pomegranate tree has tendancy of throw out suckers and as such it is rather difficult to train the plant to a single stem. Secondly, single stem training method has its own advantages. The crop is highly sucesptible to stem borer and as such training on single stem is found to be hazardous. Allowing too many stems also hinders the inter-culture operations and therefore, maintenance of 3-4 stems per plant is recommended.

Pruning

Pomegranate does not usually require pruning except for removal of suckers (water sprouts), dead and diseased branches and developing a sound framework of the tree. It is essential to remove the suckers as soon as they arise and one of the main causes of poor bearing in the commercial orchards is the failure to remove shoots in time.

The fruits are born terminally on short spurs produced all along the slow growing mature wood. These bears fruits for 3 to 4 years. Only a limited pruning of the bearing tree is therefore, required (Dutta, 1965) suggested that annual pruning in winter during dormant period should be confined to shortening of the previous seasons growth to encourage fruiting.

For securing tree and good crop, a set of new shoots should be allowed to develop every year on all sides of the tree and gradual growth of new shoots should be encouraged by restricted cutting back of the bearing shoots. It should be born in mind that shortering of the older wood is likely to reduce the yield.

There are two methods of training the pomegranate *viz.*, single stem and multi stem training. The young plants are properly trained to form a single stem with a number of well distributed scaffold limbs. Single stem is achieved by removing all the side shoots at the time of planting on this main stem, 3-4 main branches are allow to grow about 60-70 cm from the ground level. The second is much stem system in which naturally arising stems are kept. The tree in this system becomes crowdy.

Pomegranate requires less pruning except for the removal of suckers and giving a shape to the tree. The side branches are pruned and the tree is allowed to grown on a clean main stem. The fruits are born terminally on short branches know as spurs. The older spurs which have lost the capacity of fruit bearing and also those shoots which interfere with others should be removed. It is desireable to encourage new growth on 1 to 3 years old wood.

To avoid damage due to stem borer, multistem training with 4-5 stem per hiss was found beneficial. Removal of water shoots, cross branches, dead and diseased portions and suckers necessary. Pruning of old spurs (short branches) to encourage growth of new ones is suggested (Patil and Karale, 1985). Some useful tips on pomegranate pruning are:

1. Fruitful and differentiated buds are located at the distal portion of the branches.
2. Pruning of terminal portion of branch would lower down the total flower production.
3. Pruning does not affect sex ratio and fruit quality of pomegranate.
4. Pruning significantly affects total fruits, marketable and unmarketable fruits. Fruit size and yield of higher grade fruits is more with intensity of pruning.
5. Pruning minimizes the bending of branches and also the staking. Based on estimated money retums, 40 cm pruning of main stem was the best (Pawar, 1993).

Contribution of Plant Nutrients in Enhancement of Pomegranate Productivity

Pomegranate being a long duration and huge biomass (fruit) accumulating crop, removes substantial amount of plant nutrients from the soil. On the other hand pomegranate is grown on semi arid region and marginal land. Indian soils are universally deficient in nitrogen except in some parts of North Eastern region, Nearly 50 per cent soils are deficient in phosphorus and 20 per cent in potassium. Sulphur has become critical in low organic matter, coarse textured soil. There exists a huge regional disparity in fertilizer use and the consumption of plant nutrients in some of the states is far below the national average, apart from wide ratio in their use. All these point out to greater opportunity for using more balanced fertilizers for enhancing pomegranate yield, improving produce quality and maintaining system sustainability.

Higher pomegranate production with improved quality traits needs higher amounts of plant nutrients. Sugar content in pomegranate fruit is manufacture in the field and not in processing factory or sugar is synthesized in field and recovered in factory clearly brings out this fact. As such the plant nutrients through fertilizers amount for loin's share among the external production inputs. The management of plant nutrients plays a key role in influencing the number of flowers, fruits, fruit weight, juice per cent and quality parameters of pomegranate.

Nitrogen nutrition is of paramount importance for higher pomegranate productivity. Phosphate compounds in plants are called energy currency. Deficiency of phosphorus severely reduces inorganic phosphorus concentration, the photosynthetic rate and carbohydrate translocation in plants. The phosphorus has no direct effect on maturity of crop rather it counteracts the adverse effect of excessive dose of nitrogen. Potassium is fundamental for synthesis and translocation of protein and carbohydrates. It is also involved in the process of carbohydrate accumulation in plants. However, Indian soils being rich in potassium, its response is inconsistent from state to state.

The role of potassium in plant water relationship is well recognized and its top dressing at later stage minimize the adverse effect of severe soil moisture stress during drought.

Manuring and Fertilization

At the time of planting, well rotted farmyard manure @ 20 kg per tree is usually applied to pomegranate plants and this is continuously given to each plant every year at the break of monsoon. For 4-6 years old bearig, 500 g N, 250 g P_2O_5 and 250 g K_2O per plant have been found to gave good quality yield. However, 50 kg farmyard manure and 3-5 kg oil cake or 1 kg ammonium sulphate prior to flowering are recommended for healthy growth as well as fruiting. Apply the farmyard manure and half the dose of ammonium sulphate at the time of bahar treatment. The remaining dose of ammonium sulphate should be given one month after the application of the first dose.

Apart from using organic manure in the pit before planting, about 20 kg farmyard manure should be incorporated around a bearing tree at the onset of monsoon. Application of 20 kg FYM and ash per plant at the beginning of monsoon was suggested by Cheema *et al.* (1954). Phadnis (1974) recommended application of 50 kg FYM and 3-5 kg oil cake or 1 kg ammonium sulphate prior to flowering for healthy growth as well as fruiting. Experiments carried out at Mahatma Phule Krishi Vidyapeeth, Rahuri (MS) India over a period of eight years indicated the following schedule to be optimum for growth and economic production

Quantity of Nutrients/Plant	*Age of Plants*
250 g N + 125 g P_2O_5 + 125 g K_2O	1½ to 2 years
500 g N + 125 g P_2O_5 + 125 g K_2O	2½ to 3 years
500 g N + 125 g P_2O_5 + 250 g K_2O	3½ to 4 years
625 g N + 250 g P_2O_5 + 250 g K_2O	Above 4 years

Application of 0.5 kg N and 0.25 kg P_2O_5 per tree to be the optimum under the agroclimatic conditions of Udaipur in Rajastan. Pathak and Pundir (1981) recommended application of N, P_2O_5 and K_2O at 240, 160 and 60 kg respectively, per hectare for a 12 years old orchard. In a 3 year old plantation 200 kg N, 120 kg P_2O_5 and 60 kg K_2O per hectare gave the best result (Pareek, 1981).

The manures and fertilizers are applied in the basin one meter away from the trunk at the time of bahar treatment, in split doses, in case of bearing orchard.

Table 4.2: National Studies Indicated that following Fertilizer Schedule

Age (year)	*FYM (kg)*	*Dose (Gram/tree/year)*		
		N	*P_2O_5*	*K_2O*
1.5 to 2.0	10-20	250	125	125
2.0 to 3.0	20-30	500	125	125
3.5 to 4.0	30-40	500	250	250
5 onwords	50	625	250	250

The full dose of FYM, P_2O_5 and K_2O and half dose of N should be given at bahar treatment prior to first irrigation. The remaining quantity of N can be given in three equal splits at an interval of one month after the first dose. Both macro and micro nutrients affects growth and development in pomegranate. The deficiency symptoms and leaf nutrient standards for macro nutrients like N, P, K, Ca, Mg and S have been described (Wavhal, 1986) in sand culture studies on Muscut cultivar P-16. The trial on foliar applications of B, Mn, Zn and Fe indicated beneficial results. Boron application @ 0.2 per cent was found useful to reduce cracking of fruits and improvement of fruit colour. Soil application of NPK fertilizer was useful for increasing the production fruit quality of pomegranate. Bhargava and Dander (1987) observed that most of the nutrients were stabilized in 8th pair of leaf from growing tip in pomegranate and may be taken as standard for nutritional diagnosis.

Pomegranate Fertigation

Practice of applying the field with agricultural chemicals with irrigation water is called chemigation. Chemicals like fertilizers, herbicides, insecticides, fungicides, nematicides and even growth regulators can be applied through drip with irrigation water. When fertilizers are applied to the field with irrigation water through drip is called fertigation. The chemigation is a broad term, while fertigation is a specific term.

Advantages

- ☆ Accurate and uniform application under all circumstances is possible.
- ☆ Application is restricted to wetted area where active roots are concentrated.
- ☆ Nutrients can be applied as per requirement and offers opportunity to place nutrients in the root zone of crops at critical growth stages and therefore, increase fertizlier use efficiency.
- ☆ Prevent nutrient deficiencies which can develop as limited volume is exposed by roots.
- ☆ Avoid leaf burn which is possible in conventional retard development of fertilizers.
- ☆ Crop foliage is kept dry, therefore, retard development of pathogens.

- ☆ Reduce soil compaction and mechanical damage to crops.
- ☆ Save labour and fertilizers.

Limitation

- ☆ Require additional capital investment on appliances.
- ☆ Clogging of drip system if pH and mixing are not properly managed.

Criteria for Fertigation

- ☆ Avoide corrosion, softening of plastic pipes and tubing or clogging of any component of the system.
- ☆ Should be safe for field use.
- ☆ Should increase yield or at least not decrease crop yields.
- ☆ Should be completely soluble or emulsifiable in water.
- ☆ Not react adversely with salt of irrigaton water or other chemicals in irrigation water.
- ☆ It should mix uniformly and not form precipitate into the water

Research Findings on Pomegranate Fertigation

The fertigation to pomegranate @ 75 per cent R.D. through soluble fertilizers was as good as 100 per cent R.D. through conventional fertilizers on sandy clay loam soil. Application of micronutrients increased the yiled by 9.6 per cent when applied with 75 per cent R.D. through soluble fertilizers. Conventional fertilizers are to be used, urea may be applied through drip for maximum yield of pomegranate. The fertigation to pomegranate @ 75 per cent RD through soluble fertilizers increased the yield of pomegranate significantly over conventional control. It resulted in 25 per cent fertilizer saving and 15 per cent increase in yields.

The highest yield of pomegranate obtained in 100 per cent RD through soluble fertilizers (11.88 t ha^{-1}) was at par with 70 per cent N, 80 per cent P_2O_5 and 80 per cent K_2O (11.27 t ha^{-1}) and 70 per cent N, P_2O_5 and K_2O (10.81 t ha^{-1}). However, the economic analysis showed that the fertigation dose of 70 per cent N, 80 per cent P_2O_5 and 80 per cent K_2O resulted in the highest benefit:cost ratio of 3.11.

Irrigation

Although the pomegranate tree is very drought tolerant it needs adequate moisture to produce good crops. Its summer water requirements are considered by some overseas authorities to be almost as great as those of citrus. Adequate soil moisture as fruit approaches maturity is said to reduce the susceptibility of the fruit to splittigs.

The young plants require regular irrigation for their establishment and to start new growth. In northern Indias where planting is done during spring, regular watering may be given every 7 to 10 days till the onset of monsoon. However, in other areas where planting is done durig monsoon, irrigation may be given wherever there is no rain for a week or 10 days. The bed or basin system of irrigation is usually practiced. Withhold irrigation 2.5 to 3 months before bahar treatment. The fruit cracks if there is irregularity in irrigation during fruit development and therefore, copious and regular irrigation is essential during fruiting season.

The pomegranate can withstand long periods of drought. Although not much fruit is produced under drought conditions, trees will survive for years; then if properly irrigated, they grow vigorously and produce good crops. Trees will thrive and produce an abundance of fruit under high summer rainfall conditions but the fruit tends to be soft and has poor shipping and storage quality.

To produce large crops of good quality fruit pomegranates requires about the same amount of water and frequency of application as citrus. Adequate soil moisture must be maintained throughout the growing season, particularly as harvest approches in last summer and early fall, when it helps reduce the number of splity fruit. Most orchards are irrigated under furrow system, but sprinkler and drip irrigation systems are satisfactory if properly designed. Orchards thrive under non-cultivation and semi-non cultivation systems. Weed control is difficult because at present no pre emergence herbicides are registered for use in pomegranate orchards.

Table 4.3: Water Requirement of Pomegranate Plant

Cropping Season	*Month*	*Water Requirement (litres day^{-1} plant^{-1})*
Ambe	January	17
	February	18
	March	31
	April	40
	May	44
Mrig	June	30
	July	22
	August	20
Hasta	September	20
	October	19
	November	17
	December	16

Bahar Treatment/Crop Regulation

In pomegranate, there are three main flowring seasons *viz.*, Ambe bahar (February-March), Mrig bahar (June-July) coinciding with the break of monsoon and hasth bahar (September-October). Only one bahar is taken from the tree and the season of fruiting to be adopted is mainly determined by the market factor and the availability of water. The bahar treatment consists in withholding water supply for about two month before the normal flowering season to produce fresh growth and blossoms resulting in a heavy crop yield. After two months, soil is mixed with manure and light irrigation is given.

In Maharashtra, there is a continuous flowering, but the Ambe bahar (flowering in January-February) is taken. The fruits of this bahar are harvested in the second half of May or in the first of June. The advantage in taking Ambe bahar fruit is that the climate remains dry and warm to hot during the fruit development and ripening. There is less attack of pests and diseases. Moreover, the fruits are of good quality and possess attractive red colour. The Ambe bahar crop is less susceptible to cracking than the crop of other bahars. Where the irrigation facilities are not available, the Mrig bahar is recommended.

The crop is regulated to one of the bahars (season) for better management. The fruit quality of Ambe bahar is better. Sonawane (1986) reported that staggered cropping can be induced but January and February cropping was the best in quality followed by October cropping. December cropping was the poorest.

In order to regulate flowering, watering is withheld for about two months in advance of the normal flowering season. After two months, manures or fertilizers are applied and light irrigation is given. Three to four days later, heavier irrigations at normal interval are followed. The tree readily responds to this treatment by producing new growth and blooms and bears a good crop.

Normally, the pomegranate starts fruiting in about 4 years after planting. According to seasonal changes, there are three flowering seasons in Deccan, *viz.*, June-July (Mrig bahar) coinciding with the break of monsoon; February-March (Ambe bahar) and September-October (Hasta bahar). Although pomegranate may be induced to bear fruit in any of the seasons, ordinarily, only one bahar is taken from a tree, and the season of fruiting to be adopted is mainly determined by the demands, price and the availability of water.

In the Deccan, Mrig bahar is taken where water is scarce during the hot weather and, therefore, the flowering is so forced that the maximum requirement of water falls during the rainy season or immediately after the rains when plenty of water is available. For mrig bahar, treatment of withholding irrigation from December to April-May results in sufficient suppression of growth. The land is ploughed up or a shallow digging is given in April up to a depth of 20 cm. The trees usually shed their leaves by March and remain dormant condition up to May. The manures and fertilizers are applied and the first light irrigation is given in the middle of May and thereafter, one or two light irrigations may be given before the rains set in. The trees start growth by June with the onset of rains and flowers and fruits are produced. The fruits ripen in October and the harvesting continues up to December. In dry regions of western Maharashtra, the fruits get spoiled because of black spots and cracks due to high humidity in the rainy season.

Ambe bahar is also taken in the Deccan, where enough water is available during hot weather; In this case, the fruits are available during June and July and no irrigation is given after the start of the rainy season. The trees shed their leaves by October-November when a shallow digging or ploughing is done. The manure and fertilizers are applied in the month of December-January, the first irrigation is given in January and the flowers appear within a month after irrigation. In dry regions of western Maharshtra ambia bahar has been found to be better in yield and quality of fruits compared with mrig bahar fruits. Hasta bahar, the trees, have to be made dormant during August-September. This is uncertain because of the rains which occur during this period and thereafter, not usually practiced.

For standard fruit size, it is available to maintain specific number of fruits on each plant since overloading adversely affects the size and fruit quality. The number to be maintained would depend on the age and vigour of the tree.

Table 4.4: Bahar Season, Flowering Time and Harvest Period

Sl.No.	*Bahar*	*Flowering Time*	*Period of Harvest*
1.	Mrig	June-August	November-March
2.	Hasta	October-November	February-May
3.	Ambe	January-February	June-August

Plant Protection

The most important work in commercial cultivation of pomegranate is plant protection mainly against disease and fruit insects. If fruits are not treated well every fruit will be infected. Therefore, the regular treatment is spraying with fungicides, bactericides and insecticides.

Spraying schedule for control of leaf spots and fruit spots of pomegranate caused by *Alternaria colletorrichum, Cercospora* and *Drechslera* species of fungi. Leaf and fruit spots caused by these fungi are commonly observed on pomegranate fruit crop in the major pomegranate growing of he country. Spraying schedule developed at Mahtma Phule Krishi Vidyapeeth, Rahuri includes.

1. First spray of any one the fungicides *i.e.* Captan (Captaf) 0.2 per cent, Ziram (Ziride) 0.25 per cent, Mancozeb (Dithane M-45) or Indofil M-45) 0.25 per cent, copper oxychloride (Blitox-50 or Fytolan @ 0.40 per cent) or Bordeaux mixture, 1 per cent be given at the flowering/fruit setting as prophylactic spray.
2. Then, nine sprays of the same fungicide be given at 10 days interval in *Mrig bahar*.
3. Similarly, seven sprays of the same fungicide be given at 15 days interval for *Hasta bahar* cropping and five sprays of the same fungicide at 20 days interval for Ambe bahar cropping.

For getting better control of diseases, there is a need to add or mix sticker (0.5 per cent) in spray solution in each spray.

Pomegranate grown in heavy soils in state of Maharshtra affected by complex disease probably caused by one or more reasons *i.e.* insects, pathogens or both namely shot hole borer, stem borer, root knot nematodes collar rot due to *Rhizoctonia* and root rot due to *Fusarium*. The exact cause is not known. Several patterns of leaf symptoms, and tree deterioration have been first reported by Chaudhari and Desai (1993).

Table 4.5: Disease Control of Pomegranate

Sl.No.	*Disease*	*Nature of Damage*	*Control Measures*
1.	Fruit-spot (*Alternaria alternata*)	☆ Caused by fungus ☆ Small irregular spots surrounded by greenish yellow borders on fruit ☆ Later these spots/turn into strips ☆ Seeds below the rind become brownish colour	a. Prune the dead and affected twings or branches. b. Spray Dithane M-45 or Captan (500 g in 200 litres of water) after the fruit set. Generally 3-4 sprayings are needed at 15 days intervals.
2.	Fruit rot (*Aspergillus niger*)	☆ Caused by fungus and common occurrence during the rainy season. ☆ Disease affected flowers fail to set fruits and young fruits drop. ☆ Characterized by yellow or black spots all over the fruits. Concentrated on the point of attachment of the fruit to the twing. ☆ The affected fruits remain immature, under sized, lose the luster and finally become soft and start rotting. ☆ Infected fruits giving rotting or fermented smell. ☆ Fruits become yellow in colour, pale and gets flattened. ☆ In severe infection of disease, whole peels get rotten and fungal growth is seen on the surface.	a. Prune all infected twings and fruits and burn them. b. Dithane M-45 or captan should be sprayed at an interval of 15 days. c. After fruit setting carbendazin (Bavistin) @ 0.1 per cent should be sprayed in away a whole fruit surface should be covered. d. Two to three sprays are sufficient to control this disease.
3.	Leaf spot	☆ Dull violet black or blackspots. ☆ Soon the region surrounding the spot turn yellow, these spots coaleasce to form bigger spots ☆ Later, major part of the leaf becomes spotted and the leaf falls off. ☆ Fruit show dark brown to black depressed spots, which enlarge and coalesce covering the large portion of the fruits.	a. Spraying of Bavistin or JK stein. 0.1 per cent at 20 days interval is quite effective. b. Mixing of antibiotics like Agrimycin or paushmycin with above fungicide will control both bacterial and fungal infections.

Contd...

Table 4.5–*Contd...*

Sl.No.	*Disease*	*Nature of Damage*	*Control Measures*
4.	Collefotrichum leaf and fruit spots (*Colletotrichum gloeosporioides*)	☆ On leaves small purplish black or black spots are observed. ☆ Afterwards yellow hallows are observed around the spots. ☆ These spots coalesce to form big spots. ☆ Heavily infected leaves drop off. ☆ Immature and mature infected fruits are initially brownish black or blackish, round or irregular deep situated spots are observed. ☆ The infected fruits are split open, the rotting of interior portions of the fruits is observed. ☆ Disease is in severe form, the infected fruits drop off.	a. Spraying of Mancozeb (M-45) or Indofil M-45 (0.25 per cent).
5.	Cercospora leaf and fruit spots (*Cercospora punica*)	☆ Small radish or irregular spots are formed, dark black in colour. ☆ Spots coalesce to form big spots. ☆ Heavily infected leaves drop off. ☆ On the fruits, round or irregular big spots having dark black colour are formed. ☆ Under favourable conditions, these spots are formed in large size.	a. Effectively controlled by copper oxychloride fungicide @ 0.25 per cent or Bordeax mixture 1 per cent.
6.	Dreshslera fruit spots (*Drechslera rostrata*)	☆ On fruits, small raised spots are formed. ☆ Initially, they are red in colour, afterwards they convert into dark brown in colour	a. Controlled by mancozeb (Dithan M-45 or Indofil M-45 @ 0.25 per cent.

Table 4.6

Symptoms	*Control*
☆ The main stem is supported by soil mound which favours the infecton of *Rhizoctonia* at Collar portion. ☆ The bark near collar region gets rotten and it leads to wilting from dead to trunk. ☆ Poor drainage around root zone, the infection of fusarium takes place trhough root hairs and it leads to wilting of the plant. ☆ Stops the flow of water and food to plant growth, it also produces toxins like fusaric acid. ☆ If the main stem is split the blackening of xylem and phloem tissue is seen.	a) Collar rot due to *Rhizoctonia* – first of all rotted bark should be removed. Apply Bordeaux paste over the exposed area. Bordeaux paste should be prepared @ 1 kg $CuSO_4$ + 1 kg lime + 10 litres of water. b) Root rot due to *Fusarium* – soil drenching of Carbendazin @ 1 per cent (10 g Carbendazin in 10 liters of water). Such type of mixture @ 10 litres per tree should be drenched near the main stem. c) It should be followed by use of biological control *i.e.* use of *Trichoderma* one month after carbendazim drenching. The trichoderma powder @ 25 g + 5 kg well rotted FYM should be mixed thoroughly and such mixture is inoculated in soil. Drenching of carbendazim followed by use of biological control twice in a season gives better results on pomegranate wilt.

Bacterial Blight

Bacterial blight is one of the major diseases affecting pomegranate cultivation in recent years and causing enormous losses to growers. The bacterial blight affects all above ground plant parts including flowers, leaves, twings, stems, buds, and fruits, though it is more destructive when fruits are affected. All commercially grown cultivars are susceptible to this disease. On an average the disease is reported to cause 30-50 per cent losses, under favourable environmental conditions 80-100 per cent losses are also reported. The bacterial blight is also known as nodal blight or black spot. It is also known as oily spot or *Telya* in Maharashtra.

Symptoms of Bacterial Blight

A) On Leaf

1. Initially small (2-5 mm) regular to irregular grayish black water soaked (oily) spots appear on the leaves.
2. These spots appears on the leaves. These spots increase in size and turn dark, brown to black, surrounded by a yellow hallow or a water soaked margin. One to several spots appeared on a single leaf.

B) On Twing

1. Brown to black spots starts around the nodes on the twings and branches.
2. The lesions enlarge and spread on the twing leading to girdling and cracking of the nodes.

C) On Flower and Fruit

1. On fruit first water soaked lesions are seen on pericarp, which later turn brown to black.
2. Spots enlarge and coalesce to cover large area, however, infection is restricted to the rind.
3. On spot small L and Y shaped cracks appear, and in severe cases entire fruit splits open along the lesions.

The disease may be confused with fungal spots especially *Cercospora*, one can however, diffentiate it form *Cercospora* spot with experience.

Table 4.7: Difference between Oily Spot and *Cercospora* Spots

Symptoms	*Cercospora*	*Oily Spots*
Leaf	☆ Redish brown irregular spot. ☆ Smaller in size than oily spot. ☆ Numerous (few 50 spots)	☆ Dark brown to black regular to irregular spots ☆ Large than Cercospora spots. ☆ Fewer to 14-15 (Usually 4-6 spots leaf)
Fruits	☆ Irregular dark black ☆ No cracks on spots ☆ Individual spots distinct even when numerous spots lie close to each other	☆ Irregular brown to black spots. ☆ Small L-Y shaped cracks on spots, sometimes fruit splits. ☆ Several spots Coalesce to form larger spots

Pest of Pomegranate

Occurance of various pests is now becoming a major limiting factor of growing pomegranate. Aknowledge of seasonal abundance, biology of pests of this crop is a must so as to adopt improved and integrated pest management strategies effectively. Pomegrante is attacked by a large number of pests, fruit being more vulnerable than any other parts and those which cause considerable damage.

About 86 insect species are known to attack on pomegranate in the world, of which 47 have been reported from India 53 insect species associated with pomegranate in Maharashra. However, the major pests observed during different fruiting seasons of various pomegranate growing areas are butterfly (Fruit borer), fruit sucking moths, bark eating cater pillar, shot hole borer, mealy bugs, aphids, scale insects, thrips, whitefly and mites. Nematode is also a serious problem in pomegranate among these, shot hole borer and stem borer are responsible for pomegranate wilt which is the major problem of pomegranate growers presently.

The details regard damage, biology; population dynamics and management of the important pests are given in Table 4.8.

Table 4.8

Sl.No.	Pest	Nature of Damage	Control Measures
1.	Pomegranate butter fly (Fruit borer) ☆ *Deudorix Isocrates* ☆ *D. epi jarbas* ☆ *Virachola Isocrates*	☆ Newly hatched larva bores into the fruits and feeds on the pulp of seeds. ☆ The conspicous symptoms of damage are offensive smell and excreta of caterpillars coming out of the entry holes. ☆ The affected fruits falls down and thus cause heavy yield loss. ☆ About 40-90 per cent damage has been reported in different parts of India. ☆ It can attack all the developmental stages of fruit starting from flowering to harvest of fruits.	**A) Cultural and mechanical control** ☆ Restrict to only one fruiting season in a year and remove the fruits of other seasons periodically. ☆ As far as possible avoid mrig bahar ☆ Remove all the affected fruits and destroy. ☆ Bagging of fruits with muslin cloth, paper or polythin papers with sufficient aeration provision for the fruits may be practiced in the orchards where the mealy bugs is not a problem **B) Chemical control** ☆ Application of metacid 50 EC 1 mL/Litre or carbonyl 0.2 per cent (Sevin 50 per cent WP 4 g/litre) at fortnightly interval is effective. ☆ Three sprays of dimethoate @ 0.05 per cent at monthly interval was recommended ☆ Spray any one insecticide alternately starting after the fruit set – Carbaryl 50 WDP 0.2 per cent (Zirpe, 1966,) OR Deltamethrin 0.002 per cent OR Cypermethrin 0.0075 per cent (Shewale, 1994) OR – Fenvalerate 0.005 per cent OR Deltramethrin 0.003 per cent OR – Quinotphos 0.06 per cent OR Emamectin benzoate 05SG 0.05 mL/Litre **C) Biopesticides** Spraying with Neem Seed kernel extract NSKE@ 5 per cent.

Contd...

Table 4.8–*Contd...*

Sl.No.	*Pest*	*Nature of Damage*	*Control Measures*
2.	Bark eating Catterpillar ☆ *Inderbela quadrinotata* ☆ *Inderbela tetraonis*	☆ Insect bores the bark and feed inside old trees and those not maintained well are more prone to the attack of this pest ☆ Badly infested tree have 10-12 holes and such tree will not bear fruits. ☆ Larva of this pest feeds on the bark of branches or main trunk of the trees and hide in tunnels specially at the joint of branches. ☆ The excreta alongwith chewed material are seen hanging loosely on affected part of the tree. ☆ Due to seviour incidence of the pest branches or tree may dry up. ☆ The maximum damage is found in between June to April.	**A) Cultural and mechanical control** ☆ Keeping the orchard clean and avoiding overcrowding of trees help to minimize the attack of pest. ☆ In case of infestation, the affected portion requires to be cleaned by removing all webs etc. and in each hole a swab of cotton wool soaked in carbonbi sulphide, petrol or even kerosene should be inserted and holes sealed with mud. ☆ Insert the iron spike in larval tunnels to kill the larvae. **B) Chemical control** ☆ Treat the live holes on stem with 0.01 per cent Fenvalerate OR 0.05 per cent Quinolphos OR kerosene OR petrol@ 5 mL/hole and seal the holes immediately with mud. ☆ Spraying with 1 per cent carbaryl 50 WDP or 0.08 per cent Dichlorovos or 0.08 per cent Monochrotophos on infected plants **C) Biological control** Spot application of the fungus: Beauveria bassiana @ 3.6 x 101° will effectively control the pest. Control of the pest through this fungus was reported economically as compared to the chemical control
3.	Stem borer ☆ *Bactocera refomaculata* ☆ Coelosterna spinator	☆ The caterpillar makes a hole and bores through the branches. ☆ Usually main trunk or main branches are affected. ☆ The female beetles lay eggs on the stem especially during onset of monsoon.	**A) Cultural and mechanical control** ☆ Cleaning the hole and extracting the insect by inserting a hooked wire into the hole. ☆ Detect early infestatin by periodically looking out for drying branches.

Contd...

Table 4.8–*Contd...*

Sl.No.	*Pest*	*Nature of Damage*	*Control Measures*
		☆ The grub of stem borer bore into the trunk, primary and secondary branches which shows yellowing symptoms and becomes drying which later on leads to dieback symptoms. ☆ Bored branches is exposed to the secondary infection, leading to slow death of tree.	**B) Chemical control** ☆ Examine branch and stem for holes/excreta/gummosis. If detected, inject using a disposable syring (without needle) 5-10 mL of Dichlorovos 0.25 per cent and seal the hole with mud/clay. ☆ Cut out the drying portion of the branch and swab copper oxychloride 50 per cent WP on cut end. ☆ Spray all surrounding trees with quinolphos 0.05 per cent or chloropyriphos 0.05 per cent. ☆ It is good to give a prophylactic swab during May/June on the main trunk with following mixture: carbaryl (50WP) 6 g + COC (50 per cent WP) 10 g + Sticker 1 mL + Neem oil 1 mL + water 1 litre OR Chlorpyriphos 20 EC 0.1 per cent alongwith geru @ 400 g/litre of water and copper oxychloride 2.5 g/litre of water. This application may be done in a year, preferably in June-July.
4.	Fruit sucking moth Or Fruit piercing moth *Eudocima maternal* and *O. fullonia*	☆ Fruit sucking moth feed nocturnally by penetrating variety of fruits, particularly sweeter and aromatic varieties, with their highly sclerotised, heavily armored proboscis and suck the juice from the fruits. ☆ The damaging stage of this pest is adult only. ☆ The secondary infection of micro organisms takes place through these puntures due to which fruits rot and drop down.	**A) Cultural and mechanical control** ☆ Covering of fruits with stout paper bags however, this should be avoided, wherever, there is a problem of mealy bugs ☆ Adjustment of fruiting season so as to escape the stage of maturity of fruits from the peak activity of moths. ☆ Destruction of larval hosts (wild creepers) like Rangoon creepers (*Quisqualis indica*): *Tinosporea cordifolia* and *Anamirta* spp. etc. ☆ Use of flood flash light around the garden ☆ Collection and destruction of fallen/rotten fruits. ☆ Collection of moths with the help of flame device (Mashal) during 8-11 pm at night and destroy them. This is an effective method for the control of FSM.

Contd...

Table 4.8–*Contd...*

Sl.No.	*Pest*	*Nature of Damage*	*Control Measures*
			B) Chemical control ☆ Use of repellents like citronell oil (Bosch, 1971). ☆ Use of poison baits containing Malathion 20 mL alongwith jaggery (100 g) and vinegar (6 g) per litre of water is effective against *A. janata* ☆ For the control of semi loopers of fruit sucking moths (like *A. janata*) on pomegranate, spray Dichlorvas at 0.05 per cent Spray DDVP 0.1 per cent every 3-4 days around garden. ☆ Poison baits of 0.2 per cent zinc phpsphite or Acephate or Methomyl may be used alongwith attractants like jaggery, vinegar, molasses and fruit juice. A solution of 100-200 mL may be kept in small plastic bowls and hang them on trees.
5.	Shot hole borer *Euwallacea fornicates*	☆ Maximum built up of pest is found in the rainy season, this pest is inside the tunnels. ☆ During rainy season the pest attack is increasing fast ☆ With increase in relative humidity, increase the incidence of pest	**A) Cultural and mechanical control** ☆ Keep proper distance between the plants so as to avoid crowding of trees. ☆ Avoide water logging and keep soil ranked and aerated. ☆ Infested trees should be uprooted and burnt, especially the root zone. **B) Chemical control** ☆ Drench soil around main trunk with a mixture of chlorpyriphos 20 EC 2.5 mL + Tridemorph 1 mL per litre. Use 2-3 litres of mixture per tree. After three weeks repeat with Monochrotophos 1.5 mL + carbendazin @ 5 g/litres (Verghese and Jayanthi, 2001) ☆ Soil trenching with chloropyriphos 20 EC 2 mL + Dichlorvos 75 EC 0.05 mL per litre of water @ 5-10 litres/tree If pest is sever, repeat the above drenching after a month.

Contd...

Table 4.8–*Contd...*

Sl.No.	Pest	Nature of Damage	Control Measures
			☆ If infestation is low, drench with azadiractin (0.15 per cent) 3 mL/litre around the main trunk 2-3 litres of mixture/tree with Carbendazin @ 5 g/litres. ☆ Paste application to the trunk of pomegranate plant with gery @ 400 g/litre of water and copper oxychloride 2.5 g/ litre of water. This application may be done in a year preferably in June-July
6.	Aphids *Aphis punicae*	☆ Aphid infest tender shoots, flowers and tender fruits by sucking the cell sap. ☆ The infested shoots leads to shriveling. ☆ The excrete large quantity of honeydew on which black sooty mould develops which affects photosynthesis activity adversely.	**A) Chemical control** ☆ Spray with systemic insecticides Dimethoate 0.06 per cent OR Monocrotophos 0.03 per cent OR Imidacloprid 17.8 SL 0.3 mL/litre after the initiation of pest incidence. **B) Biological control** ☆ Natural enemies *viz.*, syrphids and coccinellids can sufficiently suppress the aphids. ☆ Spraying with *Verticilium lecanii* (2 x 105 CFU/g) @ 2 g/ litre off water **C) Bio pesticides** ☆ Spray NSKE 5 per cent starting from new shoot emergence.
7.	Thrips ☆ *Rhiphiphorothrips creuentatus* ☆ *Scritothrips Dorsalis*	☆ Thrips infests leaves as well as young fruits by scrapping off the tissue and suck the cell sap. ☆ The severe infestation of leaves results leaf tip curl and drying and shedding of flowers. ☆ Thrip rasp tender fruits and big fruits causing characteristics scares on the fruits and thereby reducing the luster of the fruit and ultimately affects the quality of fruits for market and exports.	**A) Chemical control** ☆ Spraying with insecticides *viz.*, Dimethoate 0.06 per cent OR Monocrotophos 0.03 per cent OR spinosad 45 SC 0.25 mL/litre prior to flowering. ☆ On severe incidence spray Methyl oxydemeton 0.05 per cent should be repeated after the fruit set. **B) Bio pesticides** ☆ Spraying of Multineem 0.5 per cent is useful (IIHR, Ann. Rep. 1998-99). ☆ Spraying with NSKE 5 per cent will also check the incidence. **C) Biological control** ☆ Spraying with *Verticillium lecanii* (2 x 105 CFU/g) @ 2 g/ litre of water

Contd...

Table 4.8–*Contd...*

Sl.No.	*Pest*	*Nature of Damage*	*Control Measures*
8.	White Fly *Siphoninus phyllyreae*	☆ The nymphs and adults suck the cell sap from the lower surface of leaves and the honey dew excretion results in the development of sooty mould and ultimately affects the yield adversely.	**A) Mechanical control** ☆ Use of yellow sticky card sheets (Fly fix®) for attracting the white flies. **B) Chemical control** ☆ Spraying with Malathion 0.1 per cent, Monocrotophos 0.08 per cent, Dimethoate 0.08 per cent Dichlorovos 0.05 per cent or Methyl demeton 0.075 are effective against whitely (Jadhav and Ajri, 1985). ☆ Spraying of Lambda Cyhalothrin 0.0025 per cent or Deltamethrin 0.0025 per cent are also effective for controlling the whiteflies. **C) Biological control** ☆ Use of parasitoids like Aphelinid. ☆ Spraying with *Verticillium lecanii* (2 x 105 CFU/g) @ 2 g/litre of water. *V. lecannii* @ 2 g/litre OR 2 mL/litre of water is quite effective for the control of white flies on pomegranate.
9.	Scale Insects *Parasaissetia nigra*	☆ Appears on shoots, twings, stem and on fruit also The nature of damage of this pest is similar to that of mealy bugs	**A) Cultural and mechanical control** ☆ Remove the heavily infested shoots and burn them. **B) Chemical control** ☆ Use of systemic insecticides for spraying. **C) Biological control** ☆ Spraying with Verticillium lecanii (2 x 105 CFU/g) @ 2 g/litre of water.

References

Amoros, A., P. Melgarejo, J.J. Martinez, F. Herna´ndez, and M. Martinez., 2000. Characterization of the fruit of five pomegranate (*Punica granatum* L.) clones cultivated in homogeneous soils. *Options Me´diterrane´-ennes er.* A 42:129–135.

Anarinco, 2006. Pomegranate history. I Sept. 2006. www. anarainco. comlhistory. htm.

Bose, T. K. and Mandal, D. P., 1972. Rooting in Pomegranate Cutting. *Indian Horti.* 17:25-26.

California Rare Fruit Growers, 1997. *Pomegranate.* 1 Sept. 2006. <www.crfg.org/pubs/ff/pomegranate.html>.

Chavan, U.D., R.N. Adsule and S.S. Kadam, 1995. Physico-chemical properties of pomegranate rind powder. *Beverage and World* 22 (1): 36.

Cheema GS and Bhat SS., 1954. *Commercial fruit of India.* MacMillan and Co. Ltd., London.

Costa, Y. and P. Melgarejo, 2000. A study of the production costs of two pomegranate varieties grown in poor quality soils. *Options Me´diterrane ennes Ser.* A 42: 49–53.

Dave Wilson Nursery, 2005. *Pomegranate* 1 Sept. 2006. <www. davewilson.com/br40/br40_trees/br40Pomegranate.html>.

Eric W. Mercure, 2007. The Pomegranate: A New Look at the Fruit of Paradise Hort. Sci. 42(5).

Faiziev, D., 1973: The effect of spacing pomegranate on cropping in the Samgarskii massif. Temat Sbornik Nauchnykh Trudov Zonal' n NII Sadovodstva i Vinogradarstva: 52-54.

Firake, N.N., and D.B. Kumbhar, 2002. Effect of different levels of N, P and K fertigation on yield and quality of pomegranate. *J. Maharashtra Agr. Univ.* 27(2): 146–148.

Gozlekci, S. and L. Kaynak, 2000. Investigations on pollen production and quality in some standard pomegranate (*Punica granatum* L.) cultivars. *Options Me´diterrane´ennes Ser. A* 42:71–77.

Gulick, P. and Vansloten, D. H., 1984. Directory of Germplasm Collections. 6-1-Tropical and Subtropical Fruits and Tree Nuts. IPPGR. Rome.

IPGRI. 2001. Regional Report CWANA, 1999–2000. International Plant Genetic Resources Institute, Rome, Italy.

Karp, D., 2006. The pomegranate: For one and all. *Fruit Gardener* 38:8– 2.

Kutuzova, A.S., 1984: Spacing density of pomegranate in the orchard. Subtropicheskie *Kul Tury* (1): 123-127.

LaRue, J.H., 1980. Growing Pomegranates in California. DANR publication leaflet 2459.1 Sept. 2006.

Levin, G.M., 1995. Aspects of pomegranate culture in Turkmenistan. *Plant Genet. Resour. Newsl.* 102:29–31.

Mars, M., 2000. Pomegranate plant material: genetic resources and breeding, a review. *Options Mediterr Ser A* 42:55–62.

Moriguchi, T., Omura, M., Natsuta, N. and Kozaki, I., 1987. *In vitro* adventitious shoot formation from anthers of pomegranate. *Hort. Science* 22(5): 947-948.

Morton, J., 1987. Pomegranate. p. 352–355. In: *Fruits of warm climates.* Julia F. Morton, Miami, FL.

Pareek, O.P., 1981. Report for the ber, datepalm, pomegranate, fig, custard apple etc. (1976-80). All India Coordinate Fruit Improvement Project. ICAR cell III, Hissar Technical Document No 12 – 20.

Pathak, S.P. and Pundir, J.P.S., 1981. Studies on response of different dose of NPK on yield and quality of pomegranate. *Udyanika,* 4: 7-11.

Patil, A. V. and Karale, A.R., 1985. In T. K. Bose, *Fruits of India: Tropical and Subtropical* 537-48. Calcutta: Naya Prokash.

Patil, A.V., Karale, A.R., and Bose, T.K., 2002. Pomegranate. In: Bose, T.K., Mitra, S.K.,Sanyal, D. (Eds.), *Fruits: Tropical and Subtropical,* vol 2. Naya Udyog, BidhanSarani, Calcutta, India, pp. 125–162.

Patil, P. K. and Patil, V.K., 1982. Effect of Soil ESP on the Growth of Pomegranate. *J. Maharashtra Agric. Univ.* 7: 273-274

Pawar, S. K., 1993. Pruning studies in (*Punica granatum* L.) Cv. Ganesh. M.Sc. (Agri.) thesis submitted to Mahatma Phule Krishi Vidyapeeth, Rahuri (M.S.) India.

Phadnis, N.A., 1974. Pomegranate for Dessert and Juice. *Indian Horti.* 19: 9-13.

Rajanna, N., 1981. Studes on the promegranate of Basein seedless pomegranate (*Punica granatun* L.) by cutting. *M.Sc. (Agri.) Thesis,* University of Agricultural Sciences, Banglore.

Salunkhe D. K., Kadam, S.S., 1995. *Handbook of Fruit Science and Technology: Production, Composition, Storage, and Processing.* New York: Marcel Dekker, p. 611.

Sulladamath, U.V., 1985. Progress Report, Third National Workshop on Arid Zone Fruit Research. MPAU, Rahuri, 5-8 July, 1985.

Pomegranate Varieties

1. Ganesh

2. Maskat

3. Phule Bhagawa

Fertilizer application in Pomegranate

Mulching for moisture conservation

5
Aonla

S.M. Todmal, S.D. Kale and A.D. Kadlag

Department of Soil Science and Agricultural Chemsitry, MPKV, Rahuri, Maharashtra

There are about 150 species of human consumable minor fruits grown in India (Mazundar, 2004). Minor fruits are less popular. They are hardy in nature withstand adverse soil and climatic condition and have long protective and productive life. A number of excellent products are made from many of these fruits, many are rich source of vitamins, minerals, enzymes, proteins, fats and antioxidants (Das and Das, 2003).

Aonla also known as Indian gosseberry (*Emblica officinalis* Sys. *Phyllanthus emblica*), is indigenous to Indian subcontinent. Owing to its hardy nature, suitability to various wastelands, high productivity and nutritive ingredients besides having therapeutic values, aonla has become an important fruit. In fact, aonla, in its processed form is very popular among the social elites. The crop ranges in status from insignificant in the western world to highly prized in tropical Asia (Morton, 1987) India ranks first in the area (95,000 ha) and production (9,61,000 Tonnes) during, 2011-12 (Anonymous, 2013). Uttar Pradesh, Gujarat, Rajasthan, Tamil Nadu, Haryana, Maharashtra and Himachal Pradesh are the major aonla growing states in India. Uttar Pradesh ranks first in area and production.

Aonla is in cultivation in India since time immemorial. It is an important minor fruit crop having commercial significance. It belongs to the family *Euphorbiaceae*. The crops is quite hardy, prolific bearer and highly remunerative even without much care. The fruit is capsular (drupaceous) with a flashy exocarp. (www.agriinfo.in).

Apart from India, naturally growing trees are found in different parts of the world like Sri Lanka, Cuba, Puerto Rico, USA (Hawai and Florida), Iran, Iraq,

Pakistan, China, Malaysia, Bhutan, Thailand, Vietnam, Philippines, Trinidad, Panama and Japan (www.nhb.gov.in).

The crop is being now explored for its value added products, which has bright future prospects for export particularly to European countries.

Aonla (*Phyllanthus emblica*) a euphorbiaceous, medium height tree and posseses (2n) = 28 (Perry, 1943) and Polyploids with 2n = 98 to 104 Chromosomes (Amal and Bhaghawan, 1957) . It is referred as evergreen in tropical countries but exhibits deciduous nature in subtropics. The bark is smooth and comes off in thin strips. Leaves are simple, flowers are same in leaf exiles and starts flowering in spring (Gupta and Singh, 2010).

Table 5.1: The Statewise Important Aonla Growing Belts in India

Uttar Pradesh	Pratapgarh, Raebareli, Varanasi, Banda, Jaunpur, Sultanpur, Kanpur, Agra, Mathura, Etawah, Fatehpur, Jhansi
Gujarat	Surendranagar, Bhavnagar, Jamnagar, Bhuj, Savarkantha
Rajasthan	Jaipur, Alwar
Maharashtra	Ahmednagar, Solapur, Satara, Nashik
Haryana	Reware, Gurgaon
Tamil Nadu	Dharmapuri
Andhra Pradesh	Anantpur
Karnataka	Bijapur, Belgaon, Bagalkot

Source: www.apeda.in.

Aonla is not merely a source of nutrients and medicine as well as it's cultivation is also highly remunerative for small and marginal farmers. Traditionally, aonla has been a wanton crop of forest or household, but during the last decade, unprecedented expansion has been witnessed in the sphere of aonla cultivation across the country, utilizing the wasteland. This has resulted in efficient utilization of resources culminating in increase in farmer's income, nutritional security coupled with enhanced employment and rejuvenation of wastelands.

Climate

The emblic seems to grow quality well under both arid and humid conditions. It is noted for being able to thrive in regions too dry and soil too poor for most other fruit crops (Morton, 1987).

Aonla is a subtropical fruit but its cultivation in tropical climate is quite successful. In India, it is being grown near sea coast upto 1800 m altitude. Annual rainfall of 630 mm to 800 mm is ideal for a growth. The tree is not much influenced either by hot wind or frost winter, at least upto the age of 3-4 years especially in north Indian condition. Soon after fruit set in spring the fruits remain dormant throughout summer without any growth thus, make it suitable crop for arid zone. The mature trees can tolerate freezing temperature as well as temperature as high as 46°C (Ganeshaiah, *et al.*, 1998).

Soil

Resource degradation leading to an unsuitable production system has demanded our attention for sustainable practices to assure continued production. Aonla based agri-horticultural system has immense potential to utilize and conserve rainfed area for betterment of poor farmers. Aonla being a deep - rooted deciduous tree species has a wide range of adaptability to grow in any type of soil. It is considered a highly tolerant and potential fruit species suitable for growing under salt affected and wasteland/ravine lands (Das *et al.*, 2011).

Aonla is not very particular in its soil requirement and grows well in sandy loam to clayey soils in India. However, production shall be highly benefited in deep and fertile soils. It has a great tolerance to salinity and sodicity and cultivated in pH range of 6.5 to 9.5 very successful. (Gupta, 2004). Aonla has been rated as a salt tolerant fruit crop (Pathak and Pandey, 1986). Soils with red, black with range of pH can very well accommodate aonla. The worst of soils upto 9.5 pH are also good enough to grow aonla. Prior to planting, the fields should be deeply ploughed, harrowed and leveled (Margadarshak's Agriculture, 2013).

Varieties

Most of aonla plantations have been raised from seed, therefore, exhibit great variability and the fruits are of inferior quality. There has been no standardization of aonla varieties and they are mostly known on the basis of size, colour or after the name of places where they are mainly grown. There were traditional three main varieties of aonla namely Banarasi, Francis and Chakaiya. These varieties had their own merits and demerits. In south India, fruit can be found occasionally throughout the year, suggesting the potential to modify the flowering pattern (Ram, 1982).

The salient features of some commercial varieties of Aonla trees are described in Table 5.2.

Propagation

Aonla is very fast catching ground in arid and semi – arid regions of the country. Hardiness, regular bearing habit in aonla have contributed to its popularity over other fruits crops.

Propogation by Seed

Aonla plants have been raised by seeds but the plants do not come true to type and there is a high variability. Besides, plants raised through seed are late in bearing (Gupta and Singh, 2010).

The tree is often propogated by seeds taken from over ripe fruits sun-dried to facilitate removal of the stone, or cut in half right through the stone (Morton, 1987). For raising rootstocks, seedlings are required to be raised through seeds. Seeds attain full maturity by February when they are extracted from the fruits and sown in the last week of February for setting high percentage of germination. Process of germination can be quickened by soaking dried seeds in 500 ppm GA_3 (Gibberellic acid) solution for 24 hours (Bhujbal, 1975). Also hot water treatment 80°C for 5 min

Table 5.2

Sl.No.	Name of Variety	Features
1.	Banarasi	☆ It is an oldest early maturing variety of aonla. ☆ Tree is semi-spreading tends to alternate bearing unless interplanted. ☆ Fruit maturities obtained in October – November. ☆ The fruit weight is 45-50 g per fruit. ☆ It has yellow colour and six segments which appear as joined together, its one of the characteristic features of this variety. ☆ Fruits are generally used for making *muramba* (preserve), candy etc. ☆ It starts fruiting in later stages (6-10 years tree age) and gives only 10-15 kg fruits per tree at 10 years of age and 100 to 200 kg per tree at the age of 20-30 years. ☆ Fruits are good source of Vit. C (625 mg 100 g–1) but it shows the poor keeping quality.
2.	Chakaiya	☆ The fruit size and tree size, both are smaller than Banarasi and Francis. ☆ The fruits are green in colour and slightly flat. ☆ The average fruit wt. is 30-35 g, storage life of fruit is more as compared to fruits of Banarasi and Francis. ☆ Fibre content of fruit is more than fruits of Banarasi and Francis. ☆ The smaller fruits are used for making pickles and large fruits are suitable for preserve and candy purposes. ☆ Fruiting starts at the age of 3rd years of the tree. ☆ The 10, 20 and 30 years old tree may yields 100 kg, 300 kg and 400 kg fruits per year, respectively. ☆ This cultivars is preferred over the others because of its yield (Morton, 1987). ☆ The fruit maturity is in December to mid January. ☆ Fruit are having good keeping quality and free from fruit necrosis.
3.	Francis	☆ This tree is regular producer of good crops. ☆ The branches of this variety are drooping type, therefore, it is also called Hathi Jhool. ☆ Fruit matures in mid November. ☆ Fruits are large, flatted, light green and smooth. ☆ The skin of fruit is fresh white green and yield is moderate. ☆ The fruits are prone to fruit necrosis. ☆ The fruit weight is around 40-45 g. ☆ It is an earlier bearing variety than Banarasi. ☆ A 10 year old tree yields about 50 kg fruits while at 20-30 years age it yields 300-400 kg fruits per tree. ☆ In calcareous soils of Uttar Pradesh, the fruits suffer from internal fruit necrosis during maturity. This is due to excess of calcium and deficiency of boron.

Contd...

Table 5.2–*Contd...*

Sl.No.	Name of Variety	Features
4.	Krishna (NA-5)	☆ This variety is tall and of spreading growth habit. ☆ It is high yielding and early variety of aonla. ☆ It is a seedling selection from Banarasi. ☆ The fruits are large, triangular and conical. ☆ The skin of fruit is smooth, whitish, green to apricot yellow with red spot on exposed portion and pinkish green and highly astringes. ☆ Since, its pulp is fairly free from fibers, it is highly suitable for *muramba* making (preserve). ☆ It bears moderately (41 kg $plant^{-1}$ of eight year old). ☆ Fruits matures in mid October to mid November.
5.	Kanchan (NA-4)	☆ It is a seedling selection of Chakaiya. ☆ Trees of this variety are spreading type. ☆ Fruits are of medium size with average weight 30-35 g. ☆ Its pulp is highly fibrous, therefore, preferred by industry for pulp extraction and manufacture of various products e.g. Pickles. ☆ The fruits are flattened and oblong. The skin of the fruit is thin and keeping quality is good. ☆ It bears profusely (74 kg fruit from an eight year old tree).
6.	Narendra Aonla – 6	☆ This is a selection from Chakaiya cultivar. ☆ Fruits are most attractive and shining, medium to large size, flattened and low fibre content. ☆ It has a mid season maturity (mid November – mid December). ☆ It bears heavily (56 kg from eight years old tree). ☆ Can be successfully cultivated on sodic soil (Rao *et al.*, 2009).
7.	Narendra Aonla – 7	☆ It is a seedling slection from Francis. ☆ Fruits are of medium to large size with conical apex. ☆ Fibre content is little higher than Narendra Aonla – 6. ☆ It is a mid season variety. ☆ It is precious, prolific and regular bearer. ☆ This is ideal variety for preparation of food and medicinal products of great promise.
8.	Narendra Aonla – 9	☆ Trees are tall, semi spreading and early maturing variety. ☆ Average fruit weight is about 50 g. ☆ The trees are moderate bearer.

Contd...

Table 5.2–*Contd...*

Sl.No.	*Name of Variety*	*Features*
9.	Narendra Aonla – 10	☆ This is a chance seedling selection from cultivar Banarasi. ☆ Trees are semi tall with semi spreading habit. ☆ Fruits are attractive, medium to large in size (41 g). ☆ The fruits are flattened round in shape, skin is roughly yellowish green with pink tinge. ☆ Flesh is whitish green fibre content in fruit is higher. ☆ It is an early maturing variety. ☆ It is highly productive and ideal for commercial cultivation.

hastens germination which takes about 10 days. Seedlings can be raised in seed beds or in polythene bags which takes about four months to attain bud size. The sodicity of soils adversely affects the seed germination (Tewari *et al.,* 1999).

Vegetative Propagation

In order to overcome the disadvantages of seed propogation multiplication of superior types of aonla has been suggested by adopting vegetative methods. Raising plants in polythene bag and grafting them in bags, ensures cent per cent field establishment on account of undisturbed root system (Tiwari and Bajpai, 2005). Recently, grafting technique was standardized in aonla which ensured multiplication of known variety plants within a year of seed sowing (Tiwari and Bajpai, 2002) of the various methods of propogation budding has been found to be the most practical and shield budding is the commercial method of aonla propogation under north India conditions.

One year old aonla seedlings with a girth of about one centimeter should be shield budded in early June with healthy and plump buds from new growth. Forket and patch budding also very successful. They can be done at any time from June to September. The most important factor of budding of aonla is the proper selection of mother plant which is highly fruitful and the bud is taken from such a branch which has good number of female flowers, otherwise, the plants will be unfruitful due to the presence of large number of male flowers.

Older trees or poor yielding trees can be changed into superior types by top working. For the rejuvenation of old aonla trees or orchards, T-budding or softwood grafting can be successfully practiced. For softwood grafting, seedling are raised *in situ* and headed back severely in the month of May for forcing new vigorous side shoots. On these new shoots, softwood grafting method can be employed by way of wedge grafting. For this purpose, use of new apex of rootstock and activated scion is necessary.

Planting

Prior to planting the field should be deeply ploughed, harrowed, leveled and marked appropriately.

The pits of the size 1m x 1m x 1m are opened during May – June at a distance of 8 to 10 m. They are left for 15-20 days exposing to sunlight. In each pit 3 to 4 baskets of farmyard manure (FYM) is mixed with dug soil and filled. After first shower, the plants are planted in the centre of these pits and staked properly. Immediately after planting, a light irrigation of the pit is carried out.

In areas with irrigation facilities, planting can be done in spring (February-March). In dry areas where mortality after transplanting is usually high, the seedling can be raised *in situ* at appropriate distance for budding with superior cloned. The seedling or vegetative saplings of aonla require clean weeding, soil working for about two years for successful establishment and fast growth.

Nutrition

The young plants should be given 15-20 kg of well rotten farmyard manure and the mature tree requires 40-50 kg in each year during January-February. In addition, application of 500 g nitrogen, 250 g phosphorus and 250 g potassium per plant per year is required. The half dose of nitrogen and full dose of phosphorus and potassium should be applied in the month of January-February and half dose of nitrogen in the month of July – August after setting of fruits. The plants need to be irrigated after application of fertilizers. Nitrogen increases vegetative growth and female flowers, while phosphorus increases sex ratio, initial set, fruit retention and yield, total soluble salts (TSS) and vitamin C content of the fruit. Potassic fertilizer increases the fruit retention and the quality.

The present day emphasis is on the sustaining agriculture which recommends less of chemical fertilizers and pesticides and more of biological inputs including microbial inoculants. Hence, use of beneficial soil micro-organisms which are cheaper and pollution free for increasing crop productivity has great potential in current day agriculture.

The chemical oriented technology was developed at a time when nobody could even imagine that it would have any deleterious effects on soil, water and atmosphere and also to the animal and human health. Indiscriminate use of chemical fertilizers, pesticides, weedicides etc. over the last few decades has adversely affected the soil fertility, water quality, yield and quality of the produce and increased level of resistance in pests (Kalloo, 2003). In aonla, the combined application of vermicompost and FYM resulted in the highest number of harvested fruits and vitamin C content as well as better growth (Korwar *et al.*, 2006).

Application of three-fourth dose of N, P_2O_5 and K_2O (750:475:750 $g^{-1}tree^{-1}$) along with 100 kg FYM $tree^{-1}$, *Azotobacter* + *Azospirillum* + PSB @ 100 g $tree^{-1}$ each improves soil nutrient level, increase the soil microbial population and plant uptake of soil available nutrients. Use of biofertilizers are always beneficial in integrated nutrient management for sustainable productivity and to a considerable extend in reducing the input cost (Bala *et al.*, 2011).

Foliar sprays of 0.6 per cent borax ($Na_2B_4O_7.10H_2O$) thrice in the months of September and October at about 10-15 days interval has also been recommended to control fruit necrosis which develops due to deficiency of boron.

In tree plants, the selection of correct leaves, stages of sampling, age of plants and concentration of other nutrients have made it difficult to interpret the result of foliar analysis (Walworth and Sumner, 1987). The critical concentration concept addresses the issues of deficiency and sufficiency of a specific nutrient element and does not addresses the nutritional balance. The diagnosis and recommendation integrated system (DRIS) was developed by Beaufils (1973), used the nutrient ratio a stable criteria with respect to the age of plants and position of leaf tissues, has been proved useful in the interpretation of leaf tissue analysis. In addition DRIS is based on nutrient balance and indicates not only the nutrient most likely to be limiting, but also the order, in which other nutrients are likely to become limiting and was able to diagnose plant nutrients are likely to become limiting and was able to diagnose plant nutrient needs early in the life of crops than sufficiency range method (Beverly *et al.*, 1986).

Table 5.3: Optimum Ranges of Elements Derived DRIS Techniques of Aonla Tree Orchards

Element	*Deficient*	*Low*	*Sufficient*	*High*	*Toxic*
N per cent	< 1.13	1.13 – 1.30	1.30 – 1.64	1.64 – 1.81	> 1.81
P per cent	< 0.036	0.036 – 0.054	0.054 – 0.092	0.092 – 0.11	> 0.11
K per cent	< 0.28	0.28 – 0.40	0.40 – 0.64	0.64 – 0.76	> 0.76
Zn (ppm)	< 25.0	25.0 – 32.4	32.4 – 45.9	45.9 – 52.9	> 52.9

Source: Nayak *et al.*, 2011.

Irrigation

Establishment of the aonla orchards in general, do not require irrigation in normal soils. At the time of establishment of orchard, first irrigation should be given just after manuring and fertilizers application. Irrigation should be avoided during flowering (mid March – mid April) period. In normal soils irrigation should be given at 15-20 days interval.

Aonla trees are hardy, and stand very well against drought. Therefore, hardly any irrigatin is practiced. The crop shall be benefited by giving 2-3 irrigations at the time of full bloom and fruit set (www.agriinfo.in).

The schedule of irrigation can also practiced as follows:

Plants up to 2 years of age – at 10 days interval.

Plants up to 4 years of age – at 15 days interval.

Plants after 4 years of age – at 20 days interval.

(aonla cultivation practices. blogspot.com)

After the monsoon rains, during October – December about 25-30 liters of water per day for tree through drip irrigation is given where drip irrigation facility is available.

Training and Pruning

Aonla tree does not require regular pruning, however, as per growth habit, shedding of all determinate shoots, the new growth in coming season is encouraged. Pruning in early years for giving proper shape and development of strong frame may be necessary for which tree should be trained to single stem upto the height of about 1 meter.

The plant should be trained to modified leader system. The frame work should be developed by encouraging the growth of 4 to 6 well spaced branches all around the trunk. Leaving only 4-5 well shaped branches with angle at about 0.75 meter from the ground level, other dead, diseased, week crises cross branches and suckers should be pruned off at the end of December. (*aonla cultivation practices, blogspot.com*)

The branches are brittle and judicious pruning to develop a strong framework is advocated to avoid branch breakage from heavy loads of fruits (Morton, 1987).

Mulching and Intercropping

During summer, the crop should be mulched with paddy straw or wheat straw at the base of the tree upto 15-20 cm from the trunk. (*blogsopt.com*)

Mulching with black polythene in aonla orchards shows maximum soil temperature throughout the period of mulching with sugarcane trash maintains minimum temperature followed by grass, paddy straw and rice husk (Rao and Pathak, 1998).

Mulching with organic materials should be applied thick enough (12 to 15 cm) to prevent weed growth simultaneously not stopping rain water penetration to the root area of late, plastic films have come into use for the purpose of mulching due to its inherent advantages of efficient moisture conservation, weed control and maintaining of soil temperature. The volume of plastics material to be used per unit area in much less or than that of traditional mulch material so that plastic films is easy to handle, transport and lay in the field (Singh and Mishra, 2007).

Since, aonla crop does not require any irrigation in summer due to dormancy of fruit, only scope of raising intercrop is in rainy season or in post monsoon period if irrigation facilities are available.

Cultivation of agricultural crops with aonla provides an opportunity for maximum use of land surface and helps to distribute the risk due to adverse climate. The tree canopy of aonla with sparse foliage allows filtered light and permits inter cropping even after the trees are fully grown. Intercropping not only generates an extra income, but also helps check soil erosion through ground coverage and improves the soil physico-chemical condition. Traditionally, intercropping in the interspaces of fruit orchards is practiced due to economic considerations, but only a few aonla based hortipastaral and horticultural systems are available (Awasthi *et al.*, 2009, Kumar *et al.*, 2008, Kumar *et al.*, 2009). The intercrops do not exert adverse effect on the growth and productivity to aonla. Turmeric, ginger and arbi can successfully cultivated in the interspaces to generate substantial additional income even at the sixth to eight year of the orchard (Das *et al.*, 2011). The leguminous crops such as moong, cowpea, urd etc. can be grown upto 6 years of tree age. At this age plants usually remain in non bearing stage and their root system and top growth are not much developed (Bakshi and Jasrotia, 2012).

During initial 2-3 years after rejuvenation, aonla groves persent an excellent opportunity for utilizing vacant interspaces in the orchard. Vegetables like bottle gourd, okra, cauliflower, coriander and turmeric, flowers like gladiolus and marigold have been found well suited for intercropping in rejuvenated aonla orchards.

In salt affected soil or marginal soils, intercropping of Dhaincha for a few years is beneficial for improving the physico-chemical properties of the soil (Singh and Mishra, 2007).

Inter Culturing

Inter culturing in aonla orchard is beneficial for to the management of soil in such a way that productivity is maintained and soil losses are minimum. The technique should be compatible to crop behaviour. The crop should be kept clean by periodical harrowing between lines of the trees. Basins should be kept clean. Grasses and weeds removed from the field can used as mulch near base of the tree.

Flowering and Fruiting

Aonla is evergreen plant in tropics. Under subtropical conditions like in North India. It begins to shed its determinate shoots along with leaves from November onwards and the indeterminate shoots are devoid of leaves by February – March. New determinate shoots begin to appear by the end of February and continue to emerge till first week of April (Bajpai, 1965 and Ram, 1971). In North India, flowering takes place from March to May. Whereas, it blooms in June –July and again in February – March in South India. Male flowers appear in clusters in leaf axils on lower parts of the determinate shoots followed by female flowers on same shoots. Some of the shoots bear only male flowers and those appearing late in the season do not flower. Aonla is cross pollinated fruit plant. Important vector for pollination are wind, honey bees and gravity. In order to improve fruit set and retention provision of proper pollinizer is essential. NA-6 and Krishna are suitable pollinizers for NA-7.

Aonla fruits are harvested at full maturing. Harvesting of fruits should be done in the morning by climbing up the tree and hand picking. Yield of aonla greately varies with cultivar. A grafted or budded tree starts bearing 3 to 4 years after planting, whereas, a seedling tree may take 6 to 8 years.

Insect Pests and Diseases

Aonla trees are attacked by number of insects and pathogens which significantly reduces the yield. Moreover, it also suffers from some physiological disorders.

Diseases

There are few important disease, which are very dangerous for the aonla cultivation. As fruits are perishable in nature, even slight injury can cause storage and transit diseases. Some of the diseased aonla is mainly attacked by fungi.

Physiological Disorders

The main physiological disorders of the tree are internal fruit necrosis and unfruitful necrosis.

1. Internal Fruit Necrosis

Internal necrosis has been observed in aonla fruits. Cultivar Francis appears to be highly susceptible followed by Banarasi. The symptoms start with the browning of the innermost part of mesocarpic tissue at the time of endocarp hardening in the 2nd or 3rd week of September which later extends towards the epicarp resulting into brownish black appearance of the flash. Infection has not been noticed on other aonla

Table 5.4: Insect Pests

Sl.No.	*Name of Pest*	*Nature of Damage*	*Management Practices*
1.	Aphids (*Cerciaphis emblica*)	The incidence of this pest is mainly seen from July to October with the peak period in September. Nymphs and adults suck the cell sap from tender shoots and leaves. They exude honeydew on which sooty mould develops. The infested leaves turn yellow and dry up. Presence of ants also indicates the infestation of aphids.	Spraying of 0.06 per cent Dimethoate.
2.	Mealy bugs (*Nipae cuccus vastator*)	Incidence of mealy bugs are been noticed from March to July with the peak population in April – May. Excessive desaping by nymphs results in drying and dropping of leaves and flowers which affect the growth of the tree/flowering and fruiting. The attacked new shorts are found bending and twisting with yellowing of leaves.	Spraying of Malathion 0.08 per cent in controlling mealy bug. Prune affected parts.
3.	Leaf twister (*Caloptilia acidula*)	Caterpillar roll the leaf and feeds inside it resulting in reduced photosynthetic capacity of leaves and subsequent leaf shedding	Spraying of 0.08 per cent malathion
4.	Gall caterpillar (*Betousa stylophora*)	The young caterpillars bore into the apical portion of the shoot during rainy season and make tunnel. Damage region bulges out abruptly into a gall which provides space for movement of caterpillar. It results in development of side shoots below the gall and apical growth checked.	Pruning of affected parts and spraying with 2 per cent parathion. Prophylactic sprays of systemic insecticide like Roger® 0.03 per cent is also suggested. Overcrowding of branches should be discouraged.
5.	Fruit borer (*Virachola Isocrates, Meridarchis scyrodes*)	Caterpillar bore into fruits and feed within fruits. It causes premature drop of fruits during monsoon season. Immature larvae are creamish white, while mature are pinkish – orange in colour. In the beginning, small grey-black spots appear at the site of in festation. Desi varieties of aonla are more susceptible.	Spray of 0.2 per cent carboryl or 0.05 per cent guinalphos at the beginning of the fruiting is recommended.

Contd...

Table 5.4–*Contd...*

Sl.No.	*Name of Pest*	*Nature of Damage*	*Management Practices*
6.	Bark eating caterpillar (*Inderbela terraonis moore*)	The pest is widely distributed throughout the country. Generally, the incidence of the pest is more common in neglected and poorly managed aonla orchards. The infestation of this pest starts in the month of April with the emergence of moths. The attack of this pest may be identified by the presence of irregular tunnels and patches covered with silkehweb consisting of excreta and chewed up wood particles, on the shoots, branches and trunk.	Spraying of 0.05 per cent Dimethoate or injecting kerosene/petrol in the holes and plugging them with cotton or mud during September -October and February –March. The larvae of this pest are parasitized by entomoseous fungus. *Beauveria bassiana* in nature. This can be used as a potential bio control agent.

Table 5.5: Diseases

Sl.No.	*Name of Diseases*	*Nature of Damage*	*Management Practices*
1.	Rust (*Ravenalia emblicae*)	Circular or semicircular, reddish solitary or gregarious spots are found on leaves from the beginning of August. Generally, one or two posales measuring 10 to 20 cm in diameter appear on infected fruits.	Spraying of 0.2 per cent Dithane Z-78 at the interval of 7 to 20 days during the months of July to September or Spraying of Mancozeb (0.2 per cent) **or** Bordeaux mixture 1 per cent **or** Plantrax (0.05 per cent) **or** Bitertanol (0.1 per cent) reduces the disease incidence (Kumar and Singh, 2000) .
2.	Fruit rot (*Phoma putaminum*)	This disease does not affect fruits much in orchard. The major loss takes place during transit to the market. The earliest symptoms of infection are seen as water soaked lesions on the fruit surfaces which enlarge in size, followed by development of small pin head type colonies of golden yellow colour. The older colonies turn olive green.	Spraying of Dithane Z-78 or carbendazin (0.1 per cent). It is recommended that fruit showing such symptoms are discarded. Bruising injury at the time of harvesting should be avoided as a preventive measure,fruits may also be treated with borax and salt solution.
3.	Leaf rust (*Phakospora phyllanthi*)	It is seen on leaves in the months of July and August.	Spraying of 0.1 per cent Bordeux mixture.
4.	Blue Mold (*Penicillium islandicum*)	Brown patches with water soaked areas are formed and the fruit is ultimately covered with bluish green pustules.	Hygienic storage and treatment with Borax and weak solution of sodium chloride.
5.	Wilt(*Fusarium* sp.)	Recently, wilting of aonla plants have been recorded from Rajasthan. The main reason for wilting through was attributed to frost injury, but association of *Fusarium* sp. is also a reason for it.	Mulching with available organic waste/black polythene film and cow dung pasting on tree trunk have been found quite effective in minimizing frost injury.
6.	Sooty mould	The aonla trees, which are infested with scale insect, are generally affected by sooty mould. Sooty mould causes velvety covering of black fungal growth on surface of leaves, twings and flowers.	Sooty mould can be managed by spraying starch 2 per cent.

cultivars like Chakaiya, NA-6 and NA-7. Hence, the cultivation of these varieties needs to be encouraged (Singh and Mishra, 2007).

In this disorder, necrosis symptoms start with browning of inner most part of the mesocarpic tissues at the time of endocarp hardening in the second and 3rd week of September, Browning of mesocarp extend towards the epicarp resulting in brownish black areas on the fruit surface in the second and third week of October. Depending upon the severity of the disorder, mesocarp of the affected fruit turns black and becomes corky and gummy pockets develop.

This disorder is due to the deficiency of boron. This can be controlled by spraying of zinc sulphate (0.4 per cent) + copper sulphate (0.4 per cent), 0.6 per cent Borax thrice in the month of September and October 10-15 days interval is recommended to control the disorder.

2. Unfruitfulness in Aonla

Unfruitfulness in aonla is mainly due to female flowers in the tree. On the basis of male and female flowers these bearing shoots can be divided into two types, (1) shoots bearing both male as well as female flowers, and (2) shoots bearing only male flowers. To avoid unfruitfulness in aonla can be avoided by selecting scion from the branch bearing maximum number of female flowers and aonla orchard should have the trees of two or three varieties in order to avoid unfruitfulness of a variety due to lack of female flowers.

Harvesting and Yield

A budded/grafted aonla tree starts bearing 3-4 years after planting, whereas, seedling tree may take 6 to 8 years. Vegetatively propagated plants attain full bearing after 10 to 12 years. Generally, aonla fruits are ready for harvest in November – December. The fruits are light green at first, but when they mature, the colour becomes dull greenish yellow, or rarely brick red.

As far as yield is concerned, the production varies from cultivar to cultivar. 'Banarasi' is a poor yielder as compared to Chakaiya and 'Francis'. On average, a grown up tree should yield 150 to 200 kg fruits per annum.

Storage and Marketing

Banarasi can be stored upto 6 days at ambient temperature whereas, Chakaiya and Francis can be stored upto 9 days. Under zero energy cool chamber, fruits of Chakaiya cultivar can be stored upto 16 days with minimum loss (Nath *et al.*, 1992). Fruits of Chakaiya can be stored at least 3 month in 15 per cent salt solution with minimum quality loss (Kumar *et al.*, 1992).

Most important single factor in successful marketing is the maintenance of excellent quality of the produce during distribution and it should also confirm to the required standards.

The mature fruits are hard and unyielding to the touch and so are well suited for bulk harvesting as well as distance transportation and marketing. For getting attractive price, fruits after harvest should be made in to different grades depending on the size.

Economics of Cultivation

Aonla is a very profitable crop. Since, it is the richest source of vitamin C (Gupta and Pareek, 1993), its industrial value is very high. A tree of Chakaiya variety of aonla gives 100 kg fruits at the 10 year age, about 200 kg at 20 years of tree age and 400 kg at 30 years age. Generally, 156 trees are accommodated in a hectare of land at a distance of 8m x 8m. the profit may vary from place to place and variety to variety.

Uses and Composition

Ayurveda has recommended the aonla for balanced diet and sound health and it is a important ingredient of triphala and chyavanprash. It is either in fresh form or as preserve like murambba, chutney, pickle, jam, jelly, sharbats etc., It is also used in the preparation of hair oils, hair dyes and different kinds of body lotions. (Gupta and Sing., 2010). The aonla powder is superior to synthetic vitamin C in treating deficiencies.

The fruit is also useful in haemorrahages, diarrhea, dysentery, anaemai, jaundice, dyspepsia and cough. It is also used in the preparation of ink, hairdyes, hair oils etc.

Table 5.6: The Food Value per 100 g Edible Portion (Morton, 1987)

Moisture	77.1 g
Protein	0.07 g
Fat	0.2 g
Carbohydrates	21.8 g
Fibre	1.9 g
Ash	0.5 g
Calcium	12.5 mg
Phosphorus	26.0 mg
Iron	0.48 mg
Carotene	0.01 mg
Thiamine	0.03 mg
Riboflavin	0.05 mg
Niacin	0.18 mg
Tryptophan	3.0 mg
Methionine	2.1 mg
Lysine	17.0 mg
Ascorbic Acid	625 mg

The aonla fruit is highly nutritious and one of the richest source of vitamin C. It contains fair amount of polyphenols and tannins, which retard the oxidation of vitamin C and render its value as anti-scorbatic in fresh as well as dried condition.

Fruits have fair amount of carbohydrates, vitamins and minerals like calcium, iron and phosphours. Aonla fruits are acidic, acrid, cooling, refrigerant, diuretic and laxative (Gopalan and Mohan Ram, 1996).

References

Amal, E.K.J. and Bhaghawan, R.S., 1957. Physiology and Vitamin C in *Emblica officicnalis*. *Proceeding of Indian Academy of Science* B. 47: 312 – 314.

Anonymous, 2003. *Report on State of Indian Agriculture.*

Aondcultivation Practices. Blogspot.in assessed on 16 Nov., 2013

Awasthi, O.P., Singh, I.S. and More, T.A., 2009. Performance of intercrops during establishment phase of aonla (*Emblica officnalis*) orchard. *Indian J. Agric. Sci.* 79 (8): 587 – 591.

Bajpai, P.N., 1965. Studies on Vegetative Growth and Development of Male and Female Genotypes of Aonla (*Emblica officnalis* Gaertn). *Agra Univ. J. Res. (Sci.)* 14: 167 – 186.

Bakshi Prashant and Jasrotia, Amit, 2012. Aonla: An important fruit of rainfed areas. December 18, 2012. www. Krishisandesh.com

Bala Shashi, Chaudhary, V.R. and Shukla, H.S., 2011. Effect of Organic Manure and Biofertilizers with Graded Dose of NPK on Soil and Leaf Nutrient Status of Aonla (*Emblica officnalis Gaertn.*) cv. Banarasi. *Karnataka J. Agric. Sci.* 24 (5): 709 – 711.

Beaufils, E.R., 1973. Diagnosis and Recommendation Integrated System (DRIS). A general scheme for experimentation and calibration based on principles developed from research in plant nutrition. *University of Natal Soil Science, Bulletin.* 1: 1-132.

Beverly, R.B., Summer, M.E., Letzch, W.S. and Plank, C.O., 1986. Foliar Diagnosis of Soybean by DRIS. *Commu. Soil Sci. and Plant Anal.* 17: 237 – 256.

Bhujbal, B.G., 1975. Improvement in seed germination of aonla (*Phylanthus emblica* L.). *Res. J. Maharashtra Agri. Univ.*, 6: 73 – 75.

Das, B.C. and Das, S.N., 2003. *Cultivation of Minor Fruits.* Kalyani Publishers, Kolkata, P. 138.

Das, D.K., Chaturvedi, O.P., Jha, R.K. and Rajeev Kumar, 2011. Yield, soil health and economics of aonla (*Emblica officnalis Gaertn.*) based agri horticultural systems in eastern India. *Curr. Sci.* 101 (6): 25.

Ganeshaiah, K.N., Uma Shankar R., Murali K.S., Uma Shankar, and Bawa, K.S., 1998. Extraction of non-timber forest products in the forests of Biligiri Rangan Hills, India. 5. Influence of dispersal mode on species response to anthropogenic pressure. *Econ. Bot.* 52 (3): 316 – 319.

Gopalan, I. and Mohanram, M., 1996. *Fruit*, National Institute of Nutrition. ICMR, Hyderabad, pp. 3-4.

Gupta, A.K., 2004. *Cultivation of fruits, vegetables and floriculture*, p. 37-40.

Gupta, P.K. and Singh, M.C., 2010. Modern cultivation Technology of Aonla. *Everyman's Sci.* 54 (6): 356 – 359.

Gupta, O.P. and Pareek, O.P., 1993. Dehydration of fruits. In: *Advances in Horticulture Vol. 4: Fruit Crops*. (K. L. Chand ha, O. P. Pareek; Eds.) M.P.H., New Delhi: 1931 – 1956.

Kalloo, K., 2003. Research and extension activities on organic agriculture in India. Organic farming in horticulture for sustainable production, 29-30 August, CISH, Lucknow, P. 1.

Korwar, G.R., Pratibha, G., Ravi, V. and Palinikumar, D., 2006. Influence of Organic and In-organics on Growth, Yield of aonla (*Emblica officnalis*) and Soil Quality in Semi Arid Tropics. *India J. Agric. Sci.* 76 (8): 457 – 461.

Kumar, P and Singh, H.K., 2000. Control of aonla (*Emblica officinalis Gaertn.*) rust through fungicides. *Appl. Hort Lucknow*. 2: 111-112.

Kumar, S. and Chaubey, B.K., 2008. Performance of aonla (*Emblica officnalis*) based hortipastoral system in semi-arid region under rainfed situation. *Indian J. Agric. Sci.* 78 (9): 744-751.

Kumar, S., Kumar, S. and Chaubey, B.K., 2009. Aonla based hortipastoral system for soil nutrient buildings and profitability. *Ann. Arid Zone*, 48: 153-157.

Kumar, S., Nath, V., Singh, I.S. and Nainwal, N.C., 1992. *Prog. Hort.* 24 (1-2): 83 – 86.

Margadarshak's Agriculture, www.jpn-virtualvarsity.org. assessed on 16 Nov., 2013.

Mazundar, B.C., 2004. *Minor Fruits Crops of India: Tropical and Sub Tropical*. Daya Publishing House, New Delhi, p. 145.

Morton, J.F., 1987. In: *Fruits of Warm Climates*, (Ed. Julia F. Morton), Creative Resource System, Winterville, Fl. PP. 125 – 129.

Nath, V., Singh, I.S. and Kumar, S., 1992. *Narendra Deva J. Agric. Res.* 7 (1): 121 – 126.

Nayak, A.K. and Sharma, D.K., Singh, C.S., Mishra, V.K., Singh Gurbachan and Swarup Anand, 2011. Diagnosis and Recommendation Integrated System Approach for N, P, K and Zn foliar Diagnostic Norms for Aonla in Central Indo-Gangetic Plains. *Plant Nutr.*, 34: 547 – 556.

Pathak, R.K. and Pandey, S.D., 1986. Problems and possibility of aonla (*Emblica officnalis Gaertn.*) cultivation in waste land, In: *Proc. Symposium on Plantation Opportunities in India*. Vol. II. PP. 305-306. Oxford and IBH Pub. Co.Pvt. Ltd., New Delhi.

Perry, A.P., 1943. Chromosome number and phytogenetic relationship in the *Euphorbiaceae*. *Am. J. Bot.* 30: 527 – 541.

Ram, S., 1971. *Ph.D. Thesis*, Kanpur University, Kanpur.

Ram, S.C., 1982. Introduction of off season flowering in aonla (*Emblica officinalis* Gaerth) trees. *Progu. Hort.* 14, 180 – 183.

Rao V.K. and Pathak, R.K., 1998. Effect of mulching treatments on growth, water relations and fruit yield of sapota (*Acharus sapota*). *Indian Agril. Sci.*, 68 (10): 657 – 660.

Rao, V.K., Rathore, A.C. and Singh, H.K., 2009. Screening of aonla (*Emblica officnalis Gaertn.*) cultivars for leaf chlorophyll and amino acid under different sodicity and salinity levels. *Indian J. Soil Cons.*, 37 (3): 193-196. 2009.

Singh Gorakh and Mishra Rajneesh, 2007. Aonla Rejuvenation, PFDC, CISH, Lucknow, PP – 16.

Tiwari, R.K. and Bajpai, C.K., 2002. Propagation of aonla (*Emblica officinalis* Gaerth) through grafting in polybags. *Indian J. Agric. Sci.* 72 (6): 353 – 354.

Tiwari, R.K. and Bajpai, C.K., 2005. Bench Grafting in Aonla (*Emblica officinalis* Gaertn.) for Easy Transport and Establishment. *Indian J. Agroforestry.* 7 (1): 65–67.

Tewari, R.K., Pandey, D. and Pathak, R.K., 1999. Effect of sodicity on the seed germination of aonla (*Emblica officinalis* Gaerth). *The Hort. J.* 12 (1): 77-80.

Walworth, J.L. and Sumner, M.E., 1987. The diagnosis and recommendation integrated system (DRIS). *Adv. Soil Sci.* 6: 149–188.

www. Blogspot.com, aonla cultivation practices, assessed on 8 Nov., 2013.

www.agriinfo.in assessed on 8 Nov., 2013.

www.apeda.in assessed on 8 Nov., 2013.

www.nbh.gov.in assessed on 16 Nov., 2013.

Aonla tree

Aonla orchard

Aonla fruit

Aonla fruit

6

Custard Apple

S.D. Kale, S.M. Todmal and A.D. Kadlag

Department of Soil Science and Agril. Chemistry, MPKV, Rahuri, Maharashtra

Custard apple (*Annona sqnomosa* L.) is one of the finest introduction in India from West Indies and South America (Bailey, 1949). It is cultivated in Andhra Pradesh, Maharashtra, Karnataka, Bihar, Orissa, Assam and Tamil Nadu. Besides it is common in China, Phillipines, Egypt and Central Africa. Number of fruits are still to be exploited to its maximum extent. The custard apple (*Annona sqnomosa* L.) is one of them with 20497 MT productions from 4990 ha of area in Maharashtra (Sontakke, 2003). It is exported in large quantity from India to UAE, Saudi Arabia, Bangladesh and Kuwait (Chanda, 1995).

Custard apple grows very well even on shallow soils and leaving it to the nature, its flowering coincides with maximum availability period. It sheds off leaves during stress period to evade moisture loss from plant tissue through transpiration and thus, is a most appropriate fruit crop for arid region. Under tropical condition, if water is available it can be made to flower anytime of the year to obtain off season fruits. The crop is known by varied names like sitaphal, sugar apple, sweet sop and sarifa. Custard apple has recently become very popular in certain areas of India, especially in Maharashtra and Gujarat. Nutritionally it is a very rich fruit and its organized orcharding is more profitable than growing cereal crops.

Custard apple is considered as one of the most delicious fruits meant for Table purpose. It is also used in ice creams and preserves as jam, jelly or other products on limited scale. The edible portion of fruit or pulp is creamy with good blend of sweetness and acidity. The leaves are known to have insecticidal properties. The nutritional value of the fruit has been given in the Table 6.1.

Table 6.1: Composition of Custard Apple Fruits

Constituents	*Content (g 100⁻ᵍ of edible portion)*
Moisture	73.5
Carbohydrates	23.9
Protein	1.6
Fat	0.3
Calcium	0.02
Phosphorus	0.04
Iron	1.0 mg
Vitamin C	37 mg

Soil

Custard apple is woody shrub with shallow root system, hence tolerant to any type of soil except alkalinity and water logged soils. Well drained soils with pH range from 5.5 to 6.5 are most preferred. Little salinity or acidity does not affect productivity but alkalinity, poor drainage, marshy wet land hamper the growth and fruiting.

Custard apple is the best suited to sandy loam soils with normal pH value. Light soils having gravels and small pebble is also suitable for its cultivation. In Andhra Pradesh and Deccan plateaue region custard apple trees are found in wide type of soil condition. Heavy soils with clay content for its cultivation is not suitable.

Climate

Custard apple is tropical plant can be grown in mild sub tropic, prefers dry climate, can withstand mild frost but the fruits become hard and do not ripen in cold weather. Dry climate is suitable for flowering but the fruit set is aides by high humidity which begins with onset of mansoon. Annual rainfall of 500-750 mm is optimum. Although custard apple withstand heat and drought conditions and drought better and make it an ideal plant for arid horticulture. In India it does well from the plains upto an elevation of 4000 ft. (1220 m), in cyclone it cannot be grown above 3000 ft. (915 m).

Custard apple requires hot dry climate during flowering and high humidity at fruit setting. Flowering comes during hot dry climate of May but fruit setting takes place on onset of mansoon. Low humidity are harmful for pollination and fertilization. The plants withstands drought condition cloudy weather and also when the temperature goes below 15°C.

Temperature affects custard apple in many ways, including influencing timing and reliability of flowering, fruit growth, ripening and fruit quality. Optimum temperature for fruit development is 25°C (±5°C) and is susceptible to skin discoloration and splitting when prolonged temperature below about 13°C during the later stage of fruit development. A relative humidity of 70 to 80 per cent is the best for fruit set and development of good fruit shape.

Fruit set is also enhanced by high diurnal temperature ranges (the difference between daytime maxima and night time minima). A low diurnal temperature range decreases the chances of male and female flower parts being open at the same time to achieve pollination.

Table 6.2: Effect of Climate on Critical Stages in Custard Apple Production

Climatic Factor	*Impact on Custard Apple Growth and Development*
Very low humidity at flowering	Desiccations of pollen and fruit failure
Heat stress	Desiccations of pollen and fruit failure
Higher maximum and minimum temperatures	Earlier maturity under higher temperature conditions
Diurnal temperature	The greater the diurnal temperature range, the greater chances for male and female flower parts being open at same time and therefore, achieving pollination and fruit set.
Wet condition for 24 hours or longer	Pollen becomes too moist, it breaks down and leads to fruit set failure.

Source: Peter Deuter, 2011.

Environmental Impact on Growth and Flowering of Custard Apple

Temperature

Both soil and air temperatures have a significant effect on growth of custard apple trees.

Soil Temperature

Soil temperature influences root and leaf growth through its effect on the rootstocks. Some cultivars widely used as rootstock because they are more adapted to low and high soil temperatures than the other *Annona* species *e.g.* Cherimoya. Good root and leaf growth can be obtained with Cherimoya rootstocks at soil temperatures between 15 - 28°C.

Air Temperature

Air temperature has pronounced influence on several growth phases.

Leaf and Shoot Growth

Even though custard apples are semi-deciduous, they are relatively sensitive to low temperatures. The air temperature drops to about 12 to 13°C at which leaf and shoot growth influenced considerably. The air temperature reached to 35°C and above considerably slows the growth. Optimum temperature for growth are considered to range from 20 to 32°C.

Flowering

Temperature of 25 to 28°C during flowering (October to February) favour good fruit set. At temperature above 28°C, custard apples produce fewer flowers but more growth and drying of flower parts increases.

Fruit Growth and Development

Fruit development (the period between fruit set and fruit maturation) takes considerably longer at lower temperatures. Low temperatures below 13°C during the later stages of fruit development cause resetting of the fruit skin and in severe cases, fruit splitting.

Soil Drainage

Trees must remain healthy for good growth and flowering. Common causes of poor tree health are root and trunk diseases such as root rot and bacterial wilt. The main feeder roots of custard apple are relatively shallow and very susceptible to root rot from water logging. A depth of at least 1 m of well drained soil is recommended to avoid damage from root rot. Similarly, bacterial wilt is worst in poorly drained soils.

Wind

Custard apple trees have relatively soft wood and often weak crotches, making them extremely susceptible to wind, drainage, particularly when carrying a full crop. Wind reduces humidity during fruit set and causes damage to the fruit skin through twinge rubbing and desiccation. A protected site using either natural or planted windbreakers is essential.

Varieties

A wide range of genetic diversity exists in custard apple from which some promising cultivars have been identified. The important cultivars of custard apple are Balanagar, Washington 98797, Washington 107005, Red sitaphal, Local sitaphal, Mammoth, Barbados, British, Guinea, Brandy and Islander. Arka sahan, is hybrid developed by Indian Institute of Horticulture Research (IIHR), Banglore. Important characteristics of major cultivars are given in Table 6.3.

Table 6.3: Custard Apple Varieties and their Characteristics

Varieties	*Fruit*	*Seed Fruit^{-1}*	*TSS*	*Acidity (per cent)*
Washington 107005	165	28	21	0.22
Washington 98787	126	48	24	0.20
Mammoth	183	14	23	0.19
Balanagar	360	43	27	0.24
Red sitaphal	231	54	25	0.21
British Guinea	181	58	25	0.31
Local	225	50	26	0.24

The heterozygosity and nature of cross pollination brings wide variation in the seedling progress.

The introduction of custard apple and inter-specific hybrids such as Washington, Island Gem, Bullock Hearts have proved promising under Indian conditions. A promising hybrid 'Arka sahan' has sweet, frangane fruit with low seed content and long shelf life.

Propagation

Custard apple can be propagated through seeds and vegetative methods which are described here. Seed is the usual means of propagation. Nevertheless the tree can be multiplied by inarching or by budding or grafting into its own seedling. In India custard apple has been commonly propagated by seeds, it is advisible to adopt vegetative method of propagation to raise true-to-type plants in the orchard.

A) Seed Propagation

Custard apple is generally grown from seeds. For arid conditions, seedlings are raised in polythene bags filled with garden soils and are planted in the field at the onset of monsoon when they attain the height of 20-25 cm. The plants establish themselves during the rainy season and can withstand unfavourable dry weather. After sometimes, seeds are directly sown in field in well prepared pit at the onset of monsoon. Treatment with GA_3 at 500 ppm to scarified seed of the annona has resulted in a high percentage of germination. Seeds treated with 100 ppm GA_3 for 24 hours germinate quickly and uniformly.

B) Vegetative Propagation

Vegetative propagation is gaining popularity now a days in custard apple, because there is a great deal of variation among the seed grown trees. The fruits are never true to type and the size gradually reduces. Among the different methods of vegetative propagation stem cutting, grafting and budding are now used for multiplication of custard apple.

Cutting

Custard apple is difficult to root fruit plant, therefore, rooting does not occur easily. Under intermittent mist, 90 per cent rooting can be obtained with NAA (Naphalic acetic acid) at 5000 ppm, when the shoots are etiolated for 15 days prior to planting.

Grafting and Budding

Custard apple root stock can be grafted or budded. Grafting has been found superior to budding and is relatively easy. Inarching seems to be the most successive method of vegetative propagation in custard apple. Side grafting is also practiced to get more advantages.

The common methods of budding practiced in custard apple are shield, patch, modified, forket and chip budding. Budding is usually done in early spring or in the autumn with bud woods obtained from dormant shoots of previous year. Seedlings are usually grafted in mid to late August when sap flow has started. By this time, the seedlings are usually about 15 months old and up to 1 meter height with stem diameter of about 10 mm. The grafted plants should be large enough to plant out by January/February.

Select grafting wood from the previous season's hardened (browned-off) growth. Each piece should contain three to six well developed buds. The wood should be the same thickness as the seedling rootstocks.

Graft the seedlings 100 to 150 mm above ground level, using either a whip or cleft graft. Alternatively, bud the seedlings using a T-bud or chip bud.

Whip Graft

To whip graft, make a sloping cut about 30 mm long on the base of the scion and on the top of the stock. This cut will need to be longer if the diameter of the wood is more than 10 mm. If the stock and scion are not same size, match the cambium layers on one side only. The cambium, which is the slightly darker layer just under the dark, lifts with the bark. Secure the scion piece to the stock with budding tape. Always apply tape from the bottom upwards to produce an overlapping pattern which sheds water away from the graft.

The whip graft is difficult for the beginner, but once mastered it is a relatively quick method of grafting. Nurseries commonly use it.

Cleft Graft

The cleft graft is the easiest grafting technique for the beginner but is suitable only for young, flexible rootstocks.

Prepare the scion piece by cutting a 30 mm long, uneven wedge on the basal end. Prepare the rootstock by cutting it off about 150 mm above soil level and making a 30 mm long cut down the centre of the stem. If the rootstock is larger in diameter than the budwood, place the cut off centre. This is then called a 'modified cleft graft'. Fit the scion so that the cambium is matched with the cambium of the stock on at least one side. Tape the union from the bottom upwards with budding tape.

T-budding

For T-budding, cut a shield bud at least 40 mm long and 10 mm wide. Select a stock at least 15 mm in diameter. Make a 'T' incision, either upright or inverted. Push the shield bud in and tape it. Do not wrap the budding tape too tightly over the bud itself. The bud should start growing after 21-28 days. Cut the tape when the bud has taken. When growth is obvious, cut the stock 100 to 150 mm above the bud. Cut the remaining stub above the after the bud has grown 150 to 200 mm high.

Chip Bud/Veneer Graft

A chip bud is where a considerable piece of wood is removed with the bud. A matching piece is cut from the rootstock and the bud inserted. Chip budding is useful where the bark of the rootstock will lift easily for T-budding.

A veneer graft is similar to a chip bud except that the scion piece used has more than one bud, usually three to four. To prepare budwood for a veneer graft, first trim it to three to four buds then remove a thin slice down one side (this should only just cut into the wood to expose the cambium on the margin of the cut). Cut the bottom of the budwood into a wedge shape. Remove a similarly, shaped piece from the rootstock, insert the budwood into this cut and tape it in place.

Veneer grafts are useful where the scion wood and rootstock are of differing diameters.

Topworking

Topworking is the process where established trees in the field are grafted over to a new scion. This is an alternative to removing trees of non performing varieties and replanting with a nursery stock.

Some sap flow is necessary to be able to lift the bark to insert grafts. Strong sap flow might retard the healing (callusing) process. Topworking in autumn allows the graft to heal (callus) before growth resumes. Topworking is much less successful when temperatures are high and the sap flow is storng. If trees must be topworked in summer, shade each plant.

Cut the tree to be reworked at a height of about 50 to 60 cm. It is the best to cut back to the two main leaders rather than a single trunk. If any side limbs (called nurse limbs) exit below this, retain them until the grafts have established. Coat the upper surface of all limbs with a white water based paint or talc/bentonite solution to avoid sunburn.

The bark graft is the first option. The number of scions used depends on the size of the limb to be grafted. Two to three scions per limb are usually sufficient to ensure a reasonable chance of a take.

Prepare the scion as per the whip graft. You can improve the percentage of success by slicing a small silver of bark from the back or side of the scion where it is to be inserted into the stock. Make a vertical cut just deep enough to expose the green cambium layer. Make a vertical cut through the bark of the stock. Leaf a flap of the bark on one side of this cut and insert the budstick. Wrap with budding tape around the limb to hold the scions in the place. Use a sealant such as grafting mastic to waterproof the graft union and the exposed cut.

Place a white opaque plastic bag over the newly grafted limbs to keep them moist and cool. Additional shade will be necessary where the topworked tree is in direct sunlight. Remove the bag and tape when the scion has to grow, usually two to three weeks after grafting.

The second option for topworking is to graft or bud new shoots that emerge from the limb. Any of the grafting or budding techniques mentioned previously are suitable here.

Soft wood grafting was found most successful in Akola, Maharashtra in terms of success (87.02 per cent), survival (73.73 per cent), shoot length (37.08 cm) and number of functional leaves (12.10 per cent) when grafted during February on 12 month old rootstock (Pawar *et al.*, 2003). Soft wood grafting in March was found to be the best method with 51 per cent graft take for custard apple under Anantpur conditions in Andhara Pradesh (Pareek and Vishal Nath, 1996).

Planting Procedure

Custard apple is a medium stature tree with upright growth habitat. Planting is done at 5 m x 5 m spacing pits of 60 cm x 60 cm x 60 cm size should be prepared and filled with FYM + top soil + fertilizers (50 g N + 50 g P_2O_5 + 25 g K_2O) mixture alongwith phorate dust before the onset of monsoon. During July – August after 1 or 2 rains, the seedlings are planted in the centre of the pit.

One to two days before planting water thoroughly to wet the tree sites to a depth of 30 cm. Do not plant tree during the hottest part of the day. Follow these planting steps.

1. Dig out hole slightly deeper and wider than the bag. Do not use posthole digger or augers. Do not place fertilizer or organic materials into the hole. It is unwise to dig deep holes and fill them with subsoil. This can cause the tree to sink as the soil settles.
2. Cut the polythene bag from the plant. Examine the root ball and straighten or trim large roots sticking out at the bottom. Gently tease out the roots at the bottom of the root ball and shake away a little of the potting mix from the fibrous roots at the top of the root ball. Treat the root very gently.
3. Place the tree in the hole, positioning it so that the top of the potting mix is at ground level. Half fill the hole with soil, gently pressing the soil into contact with the root ball. Care is needed at the stage as the custard apple has a brittle root system that easily damaged. Fill hole with water. This helps to bring the soil into close contact with the root ball. Allow water to drain before completing filling.
4. Firm soil down gently with your hands (preferably do not use your feet) and leave a slight basin around the tree to hale water. Water again.
5. Mulch the trees with coarse mulch such as cereal or legume stubble 10 to 15 cm deep. Keep the mulch 10 cm away from the trunk to avoid collar rot.
6. Where perimeter windbreaks are poorly developed, use tree guard 1.5 m height. These guards also give some protection from frost, sunburn and damage from hares and wallabies. Use old fertilizer bags or cheap shade cloth around stakes can be used to form a triangular guard. Where good wind protection is available, trees can be protected from hares and wallabies by loosely wrapping the trunk with either polythene tree protector sleeves or one thickness of isolations. This also helps to protect the bark from herbicide spray drift.
7. It is good practice to stake the trees to prevent crown damage and possible infection by bacterial wilt. Drive wooden stakes into the ground outside the root ball of the tree. Place loops of strong rubber or nylon loosely around the trunk and the stakes. Maintain the staking for at least the first 12 months. Take care to remove all ties when the stakes are removed.
8. Where trees are planted in mid-summer, coat the northern and western sides of the trunk and upper parts of limbs with white water based paint or a solution of bentonite and talc in water (1:4 by weight of bentonite to talc). Both the paint and the bentonite/talc mixture can also be applied as a spray.

9. Water the tree as required two to three times a week for the next few weeks.

Interculturing

For good plant growth, the weeding should be done to keep away the weeds. Inercropping with some legumes, peas, beans and marigold flowers are commonly taken by the growers. Normally no crop is taken during winter as the plant go under rest.

Nutrition

Plant nutrition is one of the keys to achieving good orchard performance. Both deficiencies and excesses of plant nutrients can adversely affect fruit yield and quality. Fertilizer use has to be carefully managed to ensure a balanced supply of all nutrients is maintained at the critical times during leaf growth and fruit development.

Application of nutrients to custard apple depends upon the soil and climatic conditions. In light and gravelly soils, an annual dose of 50 : 50 : 25 g N : P_2O_5 : K_2O alongwith 10 kg FYM per plant was found sufficient. The doses should be increased up to 5 years of age, there after a fixed dose should be given every year. Full amount of FYM, phosphorus, potash and half of nitrogen should be applied during monsoon season and remaining amount of nitrogen should be given during spring. Besides this, application of organic manures such as castor cake and bone meal has also been found beneficial at Sangareddy and in Rayalseema region of Andhra Pradesh. Green manuring with sunhemp, moong, urd and dhaincha during monsoon period has been found effective for fertility improvement in custard apple orchards. On light soils, apply 60-80 kg manure per tree annually and they recommended the addition of nitrogen. Application of chemical fertilizer significantly increases flowering, fruit set and yield.

Generally, no manures and fertilizers are applied to rainfed crop. However, for early and bumper harvest with a good keeping quality, the fertilization is recommended to fully grown tree. A fertilizer mixture of 250g containing ammonium sulphate, single superphosphate and muriate of potash. In Gujarat, where well organized orcharding of custard apple exists, application of 10-15 kg FYM, 2-3 baskets of river silt, 500 g nitrogen, 50 g phosphorus and 50 g potash per plant is in vogue. These are applied in the beginning of monsoon. It has been observed that the plant responds better to organic matters. A study has indicated that application of 250 g nitrogen, 125 g phosphorus and 125 g potash per plant per year gave higher yield (Anonymous, 1983). Pareek and Vishal Nath (1996) reported the application of nitrogen @ 250 g per tree per year increased productivity of custard apple, while phosphorus and potassium failed to give any response. Suresh Kumar *et al.* (2009) reported the application of N at 400 g $plant^{-1}$, P_2O_5 at 250 g $plant^{-1}$ and K_2O at 400 g $plant^{-1}$ gave higher fruit yield and quality. Rao (1974) reported that malnutrition and lack of moisture increase number of stone fruits which do not ripen properly.

Why Nutrition Needs to be Carefully Managed ?

The unmanaged approach to fertilizing custard apples involves applying fertilizer throughout the season without knowing whether the soil or the plant needs it or not. This can be lead to excessively low or high levels of some nutrients in the soil and plant and can cause several problems. These includes

- ☆ Reduced yield from nutrient imbalance
- ☆ Excessive shoot vigour, resulting in a reduction in flower numbers from shading, as well as increasing the amount of pruning
- ☆ Lower fruit quality from nutrient imbalance (Woodiness and brown pulp disorder)
- ☆ Contamination of groundwater from excess nutrients being leached out of the root zone.

In addition, blanket application fails to recognize that different varieties, different blocks of trees and different soil types have different fertilizer needs. Consequently, blanket fertilizers are tending to be too much for some and too little for others.

Nutrient levels in the soil and plant need to be more carefully monitored to avoid these problems. Nutrient monitoring improves yield and fruit quality, reduces fertilizer cost and is kinder on the environment. Fertilizer application needs to be carefully managed to ensure that it is applied at appropriate times in the cycle. For nitrogen fertilizer, these including before bud breaks (to ensure rapid and uniform bud break); during the summer leaf flush (to feed the developing leaves, young fruit and main root flush) and during the main fruit development period in autumn (to help in fruit development and maturity). The quantities of fertilizer also need to be carefully managed to ensure amounts are adequate but not excessive. Too much fertilizer promotes excessive vigour, which may detrimentally affects flowering, fruit development and leaf fall. Leaf and soil analysis are essential tools in determining appropriate fertilizers.

Fertilizer Dose Based on Nutrient Replacement Concept

This based nutrient and fertilizer application rates on how much nutrient is removed by the tree, adjusted by the expected losses of nutrients through leaching and soil fixation. The starting point is determining nutrient removal by the trees. This has been calculated from research.

Depending on soil type and situation, the figures can then be adjusted for leaching by rainfall for excessive irrigation, soil fixation and soil erosion effects. The following losses could be expected.

- ☆ 30 to 50 per cent of nitrogen lost by leaching, the higher levels occurring in sandy soils
- ☆ 50 to 80 per cent of phosphorus lost by fixation, the higher levels occur is red and clayey soils

☆ 20 to 30 per cent of potassium and magnesium lost by leaching, the higher levels occurring in sandy soils.

Table 6.4: Nutrient Removed by Trees with Varying Crop Yields

Crop Yield ($t\ ha^{-1}$)	*Total Nutrient Removed ($kg\ ha^{-1}$)*				
	N	*P*	*K*	*Ca*	*Mg*
5	12.8	1.50	12.95	3.40	1.70
10	25.6	3.00	25.90	6.80	3.40
15	38.4	4.50	38.65	10.20	5.10
20	51.20	6.00	51.80	13.60	6.80
25	64.00	7.50	64.75	17.00	8.50
30	76.80	9.00	77.70	20.40	10.20

Table 6.5: Nutrient Requirement for Full Replacement (Tree nutrient removal plus adjustment for leaching and other losses)

Crop Yield ($t\ ha^{-1}$)	*Total Tree Nutrient Removed ($kg\ ha^{-1}$)*				
	N	*P*	*K*	*Ca*	*Mg*
5	16.64	3.00	16.84	3.74	2.13
10	33.28	6.00	33.67	7.48	4.25
15	49.92	9.00	50.51	11.22	6.38
20	66.56	12.00	67.34	14.96	8.50
25	83.20	15.00	84.18	18.70	10.63
30	99.84	18.00	101.10	22.44	12.75

Fertilizer Placement

Mature tree roots extend into the middle of the tree row so the whole orchard should receive some fertilizer. Spread the fertilizer under the tree canopy.

Fertigation

Fertigation (application of suitable fertilizer through the irrigation water) is recommended. It has many advantages over the manual application of solid fertilizers.

☆ It uses less labour

☆ There is more efficient nutrient uptake by the tree (fertilizers can be applied closer to the 'feeder' root)

☆ Fertilizer can be applied more regularly and conveniently.

With efficient fertilization, annual nitrogen and potassium rates can generally be reduced up to 25 per cent.

Table 6.6: The most Suitable Fertilizers for Fertigation

Fertilizer	*Main Nutrient Supplied*
Urea	Nitrogen
Calcium nitrate	Nitrogen, Calcium
Potassium nitrate	Potassium, Nitrogen
Potassium chloride	Potassium
Monoammonium Phosphate (Technical grade)	Phosphorus, Nitrogen

Several suitable commercial soluble fertilizers that supply a range of nutrients are also available; fertigation can be done every fortnight or a month is sufficient and most practical.

Table 6.7: Fertilizer Programme for a Custard Apple Orchard Yielding 25 t ha^{-1}

Nutrient	*Just Before Bud Break (August to October)*	*December*	*January*	*March*	*May*	*Total*
			Nutrient requirement (kg ha^{-1})			
Nitrogen	28	–	28	–	–	56
Phosphorus	–	–	15	–	–	15
Potassium	8	25	34	17	–	84

Apply calcium, magnesium and trace element according to leaf and soil analysis results.

Irrigation

During the initial stage of orchard establishment, frequent irrigations at 15-20 days interval should be given particularly during the summer months. Once plant established, life saving irrigations will be needed during summer months. In light soils of arid and semiarid regions, provision of water harvesting structures by proving 5 per cent inward slope to the basin around the plant and moisture conservation by mulching in the basin with black polythene or locally available organic waste should be done. If water is available, irrigation by modified basin or drip irrigation system improves its production and quality of fruits.

Custard apple does not require any irrigation and is able to produce a good crop without being irrigated provided effective water supply is ensured to rhizosphere (root zone), by developing micro-catchment in rainy season. However, in post monsoon period 2-3 irrigation help in better quality fruits and higher production. Drip irrigation is the most efficient method to manage scarce water resource which can be created by harvesting run off during monsoon in arid region. Water stress at critical times in the development of the crop can dramatically affect fruit yield, size and quality. Careful management of irrigation is a key factor in achieving good orchard performance.

In general, custard apple is grown as a rainfed crop and no irrigation given. However, for early and bumper harvest of the crop. Irrigation on flowering i.e. from May should be given till regular monsoon starts. For better flowering and fruit setting, mist sprinkling is better over flood or drip system of irrigation as it keeps to lower down the temperature and to increase in the relative humidity.

Weed Control and Mulching

Newly planted trees find it difficult to compete with weeds for water and nutrients. Weed control immediately near the young trees is vital. Weeds are the best controlled by maintaining a grassed inter row area and mulching and spot spraying around the trees. The mulched/sprayed area should be extended to just beyond the drip line of the trees, making it roughly 2 m wide. Besides reducing weeds mulching increase soil organic matter, improves soil structure and reduces fluctuations in root temperature. It also increases water retention and may reduce irrigation frequency and amount.

Mulch may be brought in or grown on site. It brought in, coarse hay or straw mulch as sorghum stubble is preferred. The grassed inter row area is a valuable source of onsite mulch. Maintaining the inter row grass sward and continue mulching and spraying for weeds as outlined under managing young trees. This keep most weeds under control and minimize the use of herbicides.

Apply a 10 to 15 cm deep layer of mulch each year after pruning and when leaf drop or defoliation is complete. Slashing from the inter row area can provide a valuable source of mulch to about 30 cm outside the edge of the tree canopy.

Use herbicides to kill any weed that grows through the mulch and along the edge of mulched area. Apply when the weeds are actively growing. Take care to prevent herbicide contact with any green part of the plant, particularly low hanging leaves. Minimize drift by using shielded, low pressure fan or flood nozzles.

Training and Pruning

Strong tree frame work in custard apple is obtained by training the young plants during the initial stage of growth. Keeping single stem upto 50-70 cm height and then including scaffold branches in all the directions makes a good frame work of tree. To avoid crowding, proper space among the branches is maintained by removing criss cross and overlapping branches. Custard apple generally bears on new growth but sometimes fruiting is also comes on old branches. Therefore, suitable combination of young and old branches should be maintained by judicious pruning. Judicious pruning to improve shape and strength of tree must be done only in spring when the sap is rising, otherwise pruning may kill the tree.

Although untrained custard apple trees will produce good yields. Fruits quality will be lower and trees will be more difficult to harvest and probably suffer limb breakage. This makes training of trees a most important operation. Training is only through pruning but other techniques such as fruit and leaf removal is also involved. When tree reach bearing age, pruning is mainly concerned with maintaining cycle of fruiting wood production and renewal. The maintenance of a strong framework continues as secondary objective.

Start pruning trees at planting or a soon a growth begins. The first pruning is the most important as it determines the basic shape of the tree. Sometimes the nursery will have pruned the trees several weeks before delivery. Thus, avoiding the need for pruning at or soon after planting.

Specific Objectives of Training and Pruning

- ✰ Produce strong, well balanced tree resistant to wind damage
- ✰ Produce an open framework that allows easy penetration of prays and access for harvesting
- ✰ Achieve an optimum balance between fruit size and the number of the fruit set
- ✰ Minimise fruit marking from limb rub
- ✰ Control tree size
- ✰ Ensure a long, productive tree life
- ✰ Produce high quality fruit free of internal blemish.

Newly planted grafts or seedlings are staked properly and trained to develop proper shape. A bearing custard apple tree is amenable to pruning and responds favourably in terms of better size of fruits. If left unpruned, it forms a bush with large number of branches. Pruning in custard apple consists of healing back and judicious thinning of weak and diseased braches which can be completed 15 days prior to expected sprouting period depending on the climatic conditions. Pruning in custard apple should not be done until the buds are ready to start growth in spring.

Growth of the tree and therefore, pruning requirement is closely linked to planting density, which in turn depends on variety, rootstock and soil type. If trees are planted at high densities, shading encourages vigorous vertical growth and increases pruning cost. It also reduces fruit size and fruit quality. Low densities improve fruit and quality but results in very large trees that increase the cost of pruning and harvesting. Yields and cash flow in the early years of the orchard are lower.

Major Pests of Custard Apple

1. Fruit borer (*Heterographis bengalella*)
2. Fruit fly (*Bactrocera zonata*)
3. Tailed mealy bug (*Ferrisa virgata*)

1. Fruit Borer (*Heterographis bengalella*)

Symptom of Damage

- ✰ Caterpillar bore into the fruits and make tunnel inside
- ✰ Feed on the internal content of the fruits
- ✰ Affected fruits fall to ground

Identification of the Pest

- ☆ Black coloured caterpillar

Management

- ☆ Collect and destroy the damaged fruits
- ☆ Spray malathion 0.1 per cent two times once at flower formation and second at fruit set

2. Fruit Fly (*Bactrocera zonata*)

Symptoms of Damage

- ☆ Maggot bore into the semi ripened fruits
- ☆ Feed on the inside fruits
- ☆ Affected fruits – shriveled, malformed, rot and fall off

Management

- ☆ Collect fallen infested fruits and dispose them by dumping in a pit and covering with soil.
- ☆ Provide summer ploughing to expose the pupa
- ☆ Immersion of fruits in hot water (45 to 47°C) for 60 minutes to kill eggs and maggots
- ☆ Monitor the activity of flies with methyl eugenol sex lure traps.
- ☆ Use bait spray combing any one of the insecticides and molasses or jaggery 10 g L^{-1},
  - Ø Malathion 50 EC 2 mL^{-1},
  - Ø Dimethoate 30 EC 2 mL^{-1},
  - Ø Carbaryl 50 WP 4 g L^{-1}.

  two sprays at 2 weeks interval before ripening of fruits.
- ☆ Prepare bait with methyl eugenol 1 per cent solution mixed with malathion 0.1 per cent.
- ☆ Take 10 mL of this mixture per trap and keep them in 25 different places in one hectare.
- ☆ Spray malathion 50 EC at 0.1 per cent or fethion 100 EC at 1 ml L^{-1}
- ☆ Field release of natural enemies Opius compensates and Spalangia Philippines

3. Tailed Mealy Bug (*Ferrisa virgata*)

Symptoms of Damage

- ☆ Adults and crawlers – setting on leaves, young shoots and fruit (between segments)

- ✰ Yellowing of leaves
- ✰ Reduction of fruit size and do not fetch premium price in the market

Management

- ✰ Collect and destroy the mealy bug infested leaves, shoots and fruits
- ✰ Spray dichlorvos 0.05 per cent, two times at new flush and shoot formation and second at fruit set
- ✰ Release Cryptolaemous montrouzieri beetles @ 10 tree^{-1}.

Integrated Pest Management for Custard Apple

Integrated pest management (IPM) approach has been globally accepted for achieving sustainability in agriculture. It has become more relevant due to number of advantages like safty to environment, pesticide free commodities, low input cost based crop production programme etc. Integrated pest management technology manages the pest population in such a manner that economic loss is avoided and adverse side effects of chemical pesticides are minimized. The Integrated pest management packages encompass for containing the pest and disease problems. Pest monitoring is one of the important components of Integrated pest management to take proper decision to pest and disease problems. Integrated pest management packages of practices for agricultural and horticultural crops will be helpful to minimize the ill effects of chemical pesticides to promote the Integrated pest management for sustainable production.

Pest Monitoring

It is an approach to analysis field situation on the following points:

1. Plant growth at different stages
2. Compensatory abilities
3. Soil condition
4. Climatic factors
5. Farmers past experience
6. Pest and defender population dynamics

This analysis of the field situation helps in decision making skill on management practices.

Integrated Pest Management Strategies

A) Cultural Practices

1. Apply FYM 250 kg ha^{-1} to improve soil health which helps in uptake of all nutrients require for its growth
2. Application of neem cake @ 450 g tree^{-1} also helps to keep plants healthy and reduce soil borne pest
3. Avoide monoculturing and adopt mixed cropping system

4. Application of castor cake and bone meal in the ratio of 2:1 in soil is beneficial for developing tolerance ability for insect pests and diseases (500 g + 250 g tree^{-1})
5. Avoide water stagnation condition for more days
6. Remove the infected plant parts and burn it
7. Two sprays of potassium include (0.5 per cent) at weekly intervals resulted higher yields.

B) Mechanical Practices

1. Pruning of branches with mealy bugs infestation.
2. Infested/diseased fruit should be collected and destroyed.
3. Solution of pesticides, Jaggery and water may be hanged 10-12 places ha^{-1} @ 0.5 litre/wide mouthed bottle for fruit fly management.

C) Biological Control

i. Conservation

Conserve potential predators *viz., Cryptolaemus montrouzieri* and *Scymnus Coceivora* for predating upon mealy bugs.

ii. Augmentation

Release 50 lady bird beetle per tree where the infestation of mealy bugs is noticed to prevent it from increasing up the population density.

iii. Biopesticides

1. Use 5 per cent NSKE for management of sucking pests.
2. Prepare mixture of 2.5 kg Trichoderma spp + 25 kg FYM, keep it for one week under shade, use 250 g mixture tree^{-1} for soil born pathogens.
3. Application of neem cake @ 500 g tree^{-1} will help to reduce soil pests and nematodes.

iv. Chemical Practices

1. Apply relatively safer pesticides if needed in the orchards which conserve effective pollinators and avoid pollen sterility.
2. Apply Bordeaux paste to infected trunks.
3. Dipping of cuttings/grafts in chloropyriphos (0.05 per cent) solution before planting helps in nematode management.

v. Weed Management

1. Weed control in initial stages of growth (Nursery stage) is necessary to get vigorously growing healthy seedling for transplanting to permanent sites at a later stage.

2. Manual method is common practice among the farmers.
3. Use of pendimethalin @ 0.75 to 1.0 kg ai ha^{-1} holds good prospect.

vi. Rodent Management

1. Adopt orchard sanitation.
2. Do not cultivate fodder crops especially oats with the orchards.
3. Make use of Bromodio/on concentrate in bait @ 0.005 per cent oil in 2 applications at the interval of a fortnight.
4. Adopt community approach.

References

Anonymous, 1983. Report, All India Coordinated Fruit Improvement Project, pp. 18-19.

Baiely, L.H., 1949. *Manual of cultivated plants*, Macmillan Co., New York. pp. 419.

Chanda, K.L., 1995. Export of fresh fruits and nuts from India. *J. Appl. Hort.* 1 (2): 1-18.

Pareek, O.P. and Vishal Nath, 1996. Coordinated Fruit Research in Indian Arid Zone: A Two decade Profile 1996-1995, pp. 67-74.

Pawar, D.M., Ingle, V.G. and Panchabhai, D. M., 2003. *Annals of Plant Physiology*. 17 (1): 53 – 55.

Peter Deuter, 2011. Climate change – Risk and Opportunity for the Custard Apple Industry. Paper presented at the National Australian Custard apple conference, 2011. Sunshine coast, Queensland. Australia. 14th -16th July, 2011

Rao, S. N., 1974. Annonas- The legendry fruit. *Indian Hort.*, 19: 19-21.

Sonatakke, M.B., 2003. Importance of rainfall fruit crops. *Winter School on Advance in Production and Management of Rainfall Fruit Crops*. pp. 1-3.

Suresh Kumar T., Girwani A., Reddy G. S. and Bhagwan A., 2009. Studies on Nutrient Management in Custard Apple Balangar. ISHS Acta Horticulture 890 : II International Symposium on Pomogranate and Minor – including Mediterranean – Fruits : ISPMM, 2009.

Vishal Nath, Dinesh Kumar and Pandey, V., 2008. *Well Versed Arid and Semi Arid Fruits*. Satish Serial Publishing House, New Delhi. Vol 1.

Custard Apple Orchard

Custard Apple Fruit

7

Tamarind

A.B. Bhosale, S.D. Kale, S.M. Todmal and A.D. Kadlag

Department of Horticulture,
MPKV, Rahuri, Maharashtra

Tamarind or *Tamarindus indica* L. of the *Fabaceae,* subfamily *Caesalpinioideae,* is an important food in the tropics. It is a multipurpose tree of which almost every part finds at least some use either nutritional or medicinal. Tamarind is indigenous to tropical Africa but it has been introduced and naturalized worldwide in over 50 countries. The major production areas are in the Asian countries India and Thailand, but also in Bangladesh, Sri Lanka, Thailand and Indonesia. In America, Mexico and Costa Rica are the biggest producers. Africa on the whole does not produce tamarind on a commercial scale, though it is widely used by the local people. Minor producing countries in Africa are Senegal, Gambia, Kenya, Tanzania and Zambia.

Food Uses

Fruits

Tamarind fruit pulp is used for seasoning, as a food component, to flavour confections, curries and sauces and is a main component in juices and certain beverages. Tamarind fruit pulp is eaten fresh and often made into a juice, infusion or brine, and can also be processed into jam and sweets. The refreshing drinks are popular in many countries around the world, though there are many different recipes. In some African countries, the juice obtained from the fruit pulp is mixed with wood ash to neutralize the sour taste of the tartaric acid. However, the most common method is to add sugar to make a pleasantly acid drink. In Ghana, the pulp is mixed with sugar and honey to make a sweet drink. Most of the producing countries manufacture drinks commercially.

Seeds

Tamarind seed is a by-product of the tamarind pulp industry. The presence of tannins and other dyeing matter in the testa make the whole seed unsuitable for direct consumption. However, the seeds become edible after soaking and boiling in water, which removes the seed coat. In the past, and even today, seeds have been wasted even though they could be ground to make a palatable livestock feed. The major industrial product of tamarind seed is the tamarind kernel powder (TKP) which is an important sizing material used in the textile, paper, and jute industries. Tamarind seed is also the raw material used in the manufacture of polysaccharide (jellose), adhesive and tannin. In 1942, two Indian scientists announced that decorticated kernels contained 46-48 per cent of a gel forming substance.

Leaves and Flowers

Tamarind leaves and flowers can be eaten as vegetables and are prepared in a variety of dishes (ICRAF, 2007). They are used to make curries, salads, stews and soups in many countries, especially in times of scarcity. Before consumption, leaves are sometimes boiled in water and prepared as tamarind fruits.

Climate

Tamarind is a weather indicating plant of tribals in Rajasthan, India. When the length of the fruits appear longer than usual, weather with abundant rains conducive to the ripening of genhu (*Triticum aestivum* L.) and chinna (*Cicer arietinum* L.) is forecasted. This is a common phenomenon throughout the forest.

Water

Tamarind grows well with an evenly distributed mean annual rainfall of 500-1500 mm. The minimum rainfall requirement is 250 mm. In areas where rainfall is low, for example in the Sahel, where annual rainfall is 300-400 mm, the trees are usually located near the water table or along water courses. The maximum annual rainfall which tamarind can tolerate is up to 4000 mm, provided that the soil is well-drained. Tamarind grows under these conditions in the wet tropics but does not flower. In fact, dry weather is important for flower initiation and if heavy rains occur during flowering tamarind does not bear fruit. Thus, in the wet tropics of northwest India and south Malaysia, the trees grow well but do not produce any fruit. Dry weather is very important for fruit development and regardless of total annual rainfall, tamarind produces more fruit when subjected to a fairly long annual dry period. Tamarind can be grown in very dry areas with supplementary irrigation and can withstand up to six months without rainfall; this is due in part to its deep and extensive root system. This can be observed in the north and south dry zones of Sri Lanka, where there is a prolonged dry season of over 4-6 months. In the intermediate zone above 300 m, a few trees can be seen, but their bearing ability is comparatively less than those grown in the dry areas.

Temperature

Essentially a tree of tropical climates, tamarind thrives under a maximum annual temperature ranging from 33-37° C to a minimum of 9.5-20° C. The trees are very sensitive to fire and frost and require protection when small. Older trees are more resistant to extremes of temperature than young trees and can withstand temperatures as high as 47° C and as low as -3 °C without serious injury. Tamarind is more cold tolerant than mango, avocado and lychee.

Light

Tamarind is a light demanding tree and grows very slowly. It is often isolated from other vegetation and the ground is usually bare around the tree due in part to the dense shading by the canopy.

Wind

The tamarind tree is very resistant to strong winds, and can tolerate violent typhoons and cyclones. Often known as the hurricane-resistant tree, it has strong and pliant branches and a deep and extensive root system, which anchors it to the ground. Another explanation for the isolated growing habit of the tree is its tolerance to harsh environments.

Altitude

Tamarind has been reported to grow from altitudes of 0-2000 m. The lack of tamarind growing at higher altitudes is associated more with the decrease in temperature than the altitude itself. Tamarind suited to 1400 m when planting a range of trees for afforestation between 1400-1600 m above sea level.

Soil

The tamarind tree can grow in a wide range of soils, no specific soil requirement. With little or no cultivation it can flourish in poor soils and on rocky terrain. In India, it tolerates sodic and saline soils where it grows in ravines and on degraded land. Older plants are more resistant to salinity than seedlings and have been found growing close to the sea. In fact, in northeast Thailand, tamarind has been reported to establish naturally in areas with recently salinised soils.

However, tamarind, thrives best in loamy, deep, well drained alluvial soil, which favours the development of a long tap root. Silt-clay and clayey soils were not condusive to plantations of tamarind in Mali but red sandy loam soils in Andhra Pradesh.

The tree does not tolerate water logging, though it also grows well in alkaline soils that its association with ant-hills and termite mounds may also be due to a preference for a slight lime content in the soil.

Propagation

Tamarind is traditionally grown from seed, although vegetative and tissue culture propagation methods have been developed to capture the attributes of specific genotypes. The methods reviewed here indicate that propagation will not be a constraint to the production of quality planting material for those interested in establishing tamarind orchards. Neither will propagation be a problem for those plant breeders and scientists wishing to select elite cultivars or to breed improved ones.

Seed Propagation

Propagation by seed is the least expensive means of propagation and can be used to produce a large number of new plants from a minimum of stock material. In addition, seed propagation is necessary in order to produce seedlings to be used as rootstocks onto which selected cultivars can be grafted. The main disadvantage of seed propagation is that seeds fail to produce plants true to type and also resulting seedlings take a longer time to produce productive plants than those produced through vegetative propagation. However, a proper understanding of the processes involved in seed germination and initial growth is an important prerequisite for the increased domestication and cultivation of tamarind.

Seed Treatments

Germination of tamarind seed can be accelerated by overcoming the hard seededness through acid or mechanical scarification and soaking in hot water. Scarification with concentrated sulphuric acid has proved to be highly effective with tamarind seed. Muhammad and Amusa (2003) observed the highest germination percentage in seed lots treated with 50 per cent sulphuric acid for 60 minutes. El-Siddig *et al.* (2001) reported that immersion of tamarind seeds in 97 per cent sulphuric acid for at least 45 minutes at 25° C is an effective method for promoting fast and synchronous germination. Scarification with sulphuric acid at 200 ml kg^{-1} for 15 minutes increased seed germination to 67.2 per cent and reduced the incidence of abnormal seedlings to 9.9 per cent and left no hard seeds. Seeds immersed in methanol, ethanol and sulphuric acid for 10 minutes produced seedlings with high vigour (Idu and Omonhinmin, 2001).

Vegetative Propagation

Vegetative propagation can be achieved by using any one of several methods. Further more, vegetative propagation is a rapid means of improving the quality of the produce to meet the varying demands of diverse commercial markets. Trees with exceptionally good fruit should be propagated vegetatively. Tamarind can be successfully propagated by root and stem cuttings or air and stem-layering or by budding and grafting.

Selection of Planting Materials

When selecting a good 'plus' tree the following points should be considered: the tree should have a good crown and strong trunk; the tree should be disease free, undamaged and have no signs of pest attack; pods and scions should be selected from a tree that is older than 15 years and is known to produce a good harvest of fruit; for consistent pods and pulp yields, it is essential that the tree produces pods every year. Some trees produce pods every two years. Once the 'plus' tree has been selected, it is necessary to prepare an area for propagation. A nursery will provide protection to the plants when very small, and a greater chance of full establishment later on.

Cuttings

Tamarind coppices well. When coppiced or young trees are damaged root suckers are produced. These can be removed and used for small scale propagation. However, the easiest and the cheapest method of vegetatively propagating of tamarind is by stem cuttings. A technique using soft wood terminal cuttings has been developed, and the protocol standardised, by the Forest Research Station at Maddimadugu, Andra Pradesh, India. Cuttings of shoots bearing new flushes of fully turgid leaves are collected in the morning. They are immediately dipped for ten seconds in 1000 ppm of indole butyric acid (IBA), and in 50 per cent isopropyl alcohol, before being planted in polypropylene tubes containing vermiculite/perlite (1:1) and placed in a mist propagator with 70-80 per cent humidity. The use of the growth regulator IBA increased the rooting of cuttings to over 94 per cent compared to the control, which recorded only 25 per cent. Furthermore, the time taken to initiate rooting was only 10-15 days with IBA, while the control took 40-50 days. Terminal cuttings have an advantage over mid stem cuttings because there is only one cut end thus, reducing the possibility of infection by disease causing organisms during or after the rooting phase.

Soft or semi-soft stem cuttings, 15-20 cm long, taken from 1-2 years old branches can also be rooted (Swaminath *et al.*, 1990). In this method, the cuttings are wrapped in moist cloth after removal from the trees to prevent moisture loss. They are then dipped in a rooting hormone, preferably IBA at 1000 ppm, and placed in a sand bed in a mist chamber. Bud initiation and root formation occurs after about 20 days and new leaves are formed after about 45 days. Hard wood cuttings will not root and hence their use should not be attempted.

Budding and Grafting

Vegetative propagation by shield and patch budding, cleft grafting, whip grafting and approach grafting are reliable methods, although they are expensive and time consuming (Purushotham and Narasimharao, 1990). Successful budding requires that the scion material has fully formed, mature, dormant buds, and that the rootstock be in a condition of active growth such that the 'bark is slipping'. In tropical areas, the bark slips anytime the plant is in active growth - which is practically year round. For patch budding, seedling rootstocks should be grown in raised beds and budded when they are 6-9 months old. This is a suitable method for large-scale multiplication of tamarind. Pathak *et al.* (1992) reported 96 per cent and 94 per cent success respectively of patch budding and modified ring budding of 9 month old seedling rootstocks. Shield and patch budding and cleft grafting have been used for producing plants for large plantations in the Philippines and Thailand.

Approach grafting is successful for difficult to graft species, even when others techniques fail, because both the scion (upper portion of the grafted pair) as well as the seedling root stock remain attached to their own root systems during the period of graft union formation. Approach grafting has been used in tamarind with up to 95 per cent success.

Time of Planting

In seasonally dry regions, the best time for field planting is at the beginning of the rainy season, as soon as there is sufficient moisture in the soil. This will reduce the need for frequent watering until the plants are firmly established in the soil. In the initial stages of growth the plants may need some watering, especially during the hot summer months. If irrigation is available, field establishment may be undertaken at any time of the year, even in the dry season. However, it is advisable to provide partial shade to the newly established plants if planting is carried out during the dry periods.

Seedling growth in the field is initially fast (about 1.2 m in the first two years) but slows later. Mean plant height increases by about 60 cm annually. Average annual growth has been reported as 50-80 cm. However, when bare rooted seedlings are transplanted initial growth may be severely checked. This can be overcome by removing some of the leaves and irrigating frequently. If, due to unforeseen circumstances, planting in the rainy season is delayed the seedlings should be cut back and the tap root trimmed. In the Philippines the planting time is from May to June and in Sri Lanka from April to June, but preferably from October to December to coincide with the onset of the main rainy season. Transplanted trees should be protected from cattle and other animals. Regular weeding around the plants is also essential for early establishment and good growth.

Pruning and Training

Initial training and pruning of young plants during the first years is essential for the development of well formed trees. Tamarind is a compact tree and produces symmetrical branches. Young trees should be pruned to allow 3-5 well-spaced branches to develop into the main scaffold structure of the tree.

In the Philippines, young trees are pruned in the early stages of growth to train 2-3 lead branches and to remove the very low branches, thus, developing a desirable frame. Bearing trees require very little pruning other than maintenance pruning to remove dead, weak and diseased branches and water sprouts. In closely planted orchards regular pruning is needed to rejuvenate fruiting wood and control the size of the trees.

Intercropping

Tamarind allows intercropping with a variety of annual crops. Vegetables and legumes can be grown during the rainy season in the interspaces in the first three to six years to augment farm income and improve soil fertility.

Once fully established, the trees will provide a regular income to the grower even when other annual crops fail at times of protected drought. The intercropping period is usually limited to four years for vegetatively propagated trees. In Thailand's central delta, intensive cropping is practised in the plantations of the sweet cultivars, which are always vegetatively propagated and transplanted. Intercropping can be extended up to even 8-10 years for the slower maturing seedling propagated plantations.

In Central America, Mexico and Brazil where there are more than 4000 hectares of well maintained tamarind orchards, intercropping is practised in the early stages. Intercrops, and where they are not used, cover crops, are useful to control competition from weeds and to conserve moisture. Intercrops should be fertilised in their own right, to reduce competition for nutrients and to maximise their yield. There is a wide choice of intercrops which can be grown depending on the soil, climate and local market demand.

Nutrition

Fertilizers and Manures

The nutritional requirement of tamarind has not yet been studied and standardised recommendations are therefore not available. Trees are known to fruit well even without fertilizer application, due to their deep and extensive root system. In experimental plots in India, 25 g urea and 25 g single superphosphate per plant have been used, although response of tamarind to fertilizer application is reported as poor. In India, chemical fertilizers are not normally applied to tamarind trees, but 5 kg of farmyard manure is applied to the planting hole at the time of planting. Every year thereafter 5 kg of farmyard manure and 5 kg of neem cake are applied per tree in the months of March and April. The commercial sweet tamarind growers in Thailand use chemical fertilizers, mostly urea @ 100-200 g tree^{-1}, which supports high yields. In the Philippines, the general recommendation is to apply 100-200 g/ tree of ammonium sulphate, about a month after planting and an equal amount at the end of the rainy season. The amount of fertilizer is gradually increased as the trees grow. When the tree begins to bear fruits, about 500g of a complete fertilizer containing high amounts of nitrogen and potassium is applied per tree twice a year. A full bearing tree should receive 2-3 kg of an NPK complete fertiliser each year, the formula depending upon the nutrient content of the soil.

Irrigation

Irrigation is not normally practised in tamarind cultivation, but promotes better growth during establishment and the early stages of growth, especially during the dry seasons. Where irrigation facilities are available watering should be done and repeated as the need arises in the early stages of growth. In later years as the deep tap root system develops, the need for watering becomes less. Flowering and fruiting is promoted by irrigation. In dry areas the use of water harvesting techniques during the rainy season should be considered as it encourages subsequent growth and fruiting. Mulching during the dry season will also help to reduce water losses from evaporation. Mulches around the trees also help in weed control and water conservation.

Harvesting Stage

In the Philippines fruits are harvested at two stages, green for flavouring and ripe for processing. The fruits of sweet types are also harvested at two stages, half-ripe and ripe. At the half ripe stage the pulp is yellowish green and has the consistency of an apple, particularly in the case of sweet forms. At the ripe stage, the pulp shrinks, due to loss of moisture and changes to reddish brown and becomes sticky.

In most countries, the sour tamarind ripe fruits are harvested by shaking the branches and the pods are collected on a mat. Sweet tamarind pods fetch a high price both in local and foreign markets and are carefully harvested by hand picking. Sometimes bamboo ladders are used to pick the fruits. If the whole fruit is marketed, both sweet and sour types should be harvested by clipping to avoid damaging the pods. The use of poles in picking is not desirable as knocking can cause damage to the pods.

Generally, the fruits are left to ripen on the tree before harvesting, so that the moisture content is reduced to about 20 per cent. Mexican studies reveal that the fruits begin to dehydrate 203 days after fruit set, losing approximately half of the moisture. When the fruits are left unharvested they may remain hanging on the tree for almost one year after flowering and eventually abscise naturally. Sometimes they remain on the tree until the next flowering period.

Yield

The yield of tamarind varies considerably in different countries, depending on genetic and environmental factors. Feungchan *et al.* (1996 a) reported that the fruit yields are influenced by environmental and genetic factors, but the age of the tree is not correlated with fruit yield. The size of the tree is obviously a factor however, and older trees which have many years to mature, would be expected to be larger and so produce a greater yield. Age of bearing is important in relation to profits. Reports of very long juvenile periods over 4-10 years are probably because they were recorded on unselected materials.

Cross-pollination results in higher fruit set and retention in tamarind than when open or self-pollinated. Bees are known to pollinate tamarind, and bee farming could provide additional income to farmers. Fruit set is influenced by environmental factors, such as age and size of shoots, carbon: nitrogen ratio and the hormonal balance under which the trees are grown. Due to the influence of these factors, high seasonal and annual yield variations can be expected.

Tree to tree variation is widely observed in growth and yield. Pod yield is sometimes cyclic, with bumper yields in every third. The tree then needs to recover for a couple of years before reaching peak performance again.

Presently, fruits are harvested from untended trees and wild types, although in Thailand a few sweet tamarind orchards have been established recently. A young tree yields 20-30 kg fruits per year, while full grown trees of 20 years of age are known to yield 150 -200 kg/tree/year or 12-16 t/ha. A full-grown tree is reported to yield about 180-225 kg of fruits per season (FAO, 1988). In India, the average production of tamarind pods per tree is 175 kg and of processed pulp is 70 kg/tree. The best yielding elite trees in Bangaluru have pod yields ranging from 250-350 kg/tree and pulp yields of 100-175 kg/tree.

Storage

Fresh fruits are often dried using small-scale dehydrators, however, in most countries rural households dry pods in the sun. The shells, fibres and seeds are then removed and the pulp stored in plastic bags or earthenware pots. The dry ripe pods can easily be cracked and the pulp and fibres separated from the broken shells. The pulp is then processed by peeling and removing the fibre strands from the pulp. After separating the pulp from the fibers, seeds and shells, it is then compressed and packed in palm leaf mats, baskets, corn husks, jute bags or plastic bags for storage and marketing. In Thailand, the pulp is mixed with salt and compressed and packed in plastic bags to exclude air for storage. Each bag weighs about 5 kg and these are repacked in boxes for transport to distant markets.

Processed Products from Tamarind

The commonly available market samples of tamarind pulp usually contains extraneous matter such as seed, dust, fiber, etc., and hence can be unfit for various culinary preparations. Value addition is of immense benefit for traders and consumers. Therefore, processed products from tamarind can meet the requirements of consumers for convenience and ready to use. Tamarind is used for the preparation of various processed products and some examples of value additions are cited below.

Tamarind Beverage

Tamarind fruit pulp is used for the preparation of beverages in different regions. Good quality ready to serve beverage, syrup and concentrate can be prepared with a shelf life of six months at ambient storage (Kotecha and Kadam, 2003 b).

Tamarind pulp was treated with pectolytic enzymes and the extract obtained was used for the preparation of flavoured RTS tamarind beverage. The carbonated tamarind beverage having 12.5 per cent juice, 16°Brix and 0.4 per cent acidity was found to be highly acceptable for up to two months of storage at room temperature.

Foam Mat Drying of Tamarind Pulp

Drying of tamarind pulp by foam mat drying using different foaming agents was reported by Vernon-Carter *et al.* (2001). The foaming agent *viz.*, mesquite gum, ovalbumin and a low molecular weight surface active blend were hydrated to 50 per cent (w/w) solutions and added to the samples either singly or in combination. It was found that the mesquite gum + surface active blend and ovalbumin + surface active blend, ovalbumin + mesquite gum resulted in acceptable flavours.

Tamarind Juice Concentrate

A process for the preparation of tamarind juice concentrate was developed by Central Food Technological Research Institute, Mysore, India. The process involves extraction of tamarind juice and concentrating the juice in a vacuum evaporator to total soluble solids of 70°Brix. The tamarind concentrate is shelf stable at room temperature and can be diluted and used in various food preparations.

Tamarind juice be concentrate be exhibited be Newtonian flow behaviour up to 19°Brix and pseudo plastic nature at higher than 23°Brix obeying the power law relationship. The flow behaviour index for concentrate of 23°Brix was 0.747 and 0.625 for concentrate of 28-62 ° Brix.

Tamarind Seed Products

Tamarind Seed Polysaccharide

Tamarind seed polysaccharide is used as a gelling, thickening or stabilising agent in foods in Japan. Tamarind kernel powder is now extensively used for sizing fabrics in textile industries. There is wider interest over the past few years in the use of tamarind seed polysaccharide extracts in the food industry.

Antioxidants from Tamarind Seed Coat

Extraction of antioxidant compounds from the seed coat of sweet Thai tamarind was reported by Luengthanaphol *et al.* (2004). The antioxidants were extracted by supercritical fluid extraction and by solvent extraction using ether or ethyl acetate. The mixture obtained by solvent extraction with ethanol showed epicatechin yield of 150mg/100g and the highest antioxidant activity in terms of peroxide value as compared to the supercritical fluid extraction.

Crude tamarind seed coat extracts were stable following heat treatment at 100°C for two hours. Antioxidative activity of the extracts was lower at pH 5.0 than at pH 3.0 or 7.0.

References

El-Siddig, K., Gunasena, H.P.M., Prasa, B.A., Pushpakumara, D.K.N.G., Ramana, K.V.R., E.W.M. Verheij and R.E. Coronel, 2001. PROSEA Foundation, Bogor. Indonesia: 298 - 301.

Feungchan, S., Yimsawat, T., Chindaprasert, S. and Kitpowsong, P., 1996a. Tamarind (*Tamaindus indica* L.) Plant Genetic Resources in Thailand. *Thai Journal of Agicultural Science*, Special Issue No. (1):1-11

FAO, 1998. Tropical Food Plants, FAO Food and Nutrition. A resource book for promoting the exploitation and consumption of food plants in arid, semi-arid and sub-humid lands of Eastern Africa. Paper 42: 478-483.

ICRAF – World Agroforestry Centre. ICRAF Agroforestry Tree Database: *Tamarindus indica* L., URL http://www.worldagroforestrycentre.org (visited on 31.01.2007).

Idu, M. and Omonhinmin, C.A., 2001. Effect of various pretreatments on the seedling growth performance of *Tamarindus indica* L. *Plant Biosystems*, 135(2):165-168.

Pathak, R.K., Ojha C.M. and Dwivedi, R., 1992. Adopt patch budding for quicker multiplication in tamarind. *Indian Horticulture*, 36(2): 17.

Kotecha, P.M. and Kadam, S.S., 2003 b. Studies on browning in tamarind pulp during storage. *Journal of Food Science and Technology* (Mysore). 40(4): 398-399.

Muhammad S. and Amusa N.A., 2003. Effects of sulphuric acid and hot water treatment on seed germination of tamarind (*Tamarindus indica* L). *African Journal of Biotechnology* Vol. 2 (9), pp. 276-279.

Purushotham K. and Narasimharao, S.B.S., 1990. Propogation of Tamarind by Veneer and Softwood Grafting. *Journal of South Indian Horticulture*, 38 (4):225

Swaminath, M.H., Ravindra, D.S., and Mumtaz, J., 1990. Propagation of tamarind through stem cuttings. *My Forest*, 26(2): 207-208.

Vernon-Carter, E.J., Espinosa-Paredes, G., Beristain, C.I. and Romero-Tehuitzil, H., 2001. Effect of foaming agents on the stability, rheological properties, drying kinetics and flavour retention of tamarind foam-mats. *Food Research International*. 34(7): 587-598.

Tamarind Tree

8
Fig

A.B. Bhosale, S.M. Todmal and S.D. Kale

Department of Horticulture,
MPKV, Rahuri, Maharashtra

The common fig (*Ficus carica*) is a species of flowering plant in the genus *Ficus,* from the family *Moraceae,* known as the common fig (or just the fig), *anjeer* (Iran, Pakistan), and *dumur* (Bengali). While the ancient history of the fig centers around the Mediterranean region and it is most commonly cultivated in mild-temperate climates, it nevertheless has its place in tropical and subtropical horticulture. Botanically identified as *Ficus carica* L. (family *Moraceae*), it is unique in a genus embracing perhaps over 1,000 species, mostly giant "rubber trees", and mostly tropical. It is almost universally known simply as fig, common fig, or edible fig. The name is very similar in French (*figue*), German (*feige*), Italian and Portuguese (*figo*). In Spanish it is *higo* or *brevo*. Haitians give it the name, *figue France,* to distinguish it from the small, dried bananas called "figs".

It is the source of the fruit also called the fig, and as such is an important crop in those areas where it is grown commercially. Native to the Middle East and Western Asia, it has been sought out and cultivated by man since ancient times, and is now widely grown throughout the temperate world, both for its fruit and as an ornamental plant.

History

The edible fig is one of the first plants that was cultivated by humans. Nine subfossil figs of a parthenocarpic (and therefore sterile) type dating to about 9400–9200 BC were found in the early Neolithic village Gilgal I (in the Jordan Valley, 13 km north of Jericho). The find predates the domestication of wheat, barley, and legumes, and may thus be the first known instance of agriculture. It is proposed that this sterile but gustatively desirable type was planted and cultivated intentionally, one thousand years before the next crops were domesticated (wheat and rye).

Figs were also a common food source for the Romans. Cato the Elder, in his *De Agri Cultura*, lists several strains of figs grown at the time he wrote his handbook: the Mariscan, African, Herculanean, Saguntine, and the black Tellanian (*De agri cultura*, ch. 8). The fruits were used, among other things, to fatten geese for the production of a precursor of foie gras.

It was cultivated from Afghanistan to Portugal, also grown in Pithoragarh in the Kumaon hills of India. From the 15th century onwards, it was grown in areas including Northern Europe and the New World. In the 16th century, Cardinal Reginald Pole introduced fig trees to Lambeth Palace in London.

Description

It is a gynodioecious (functionally dioecious), deciduous tree or large shrub. The complex inflorescence consists of a hollow fleshy structure called the syconium, which is lined with numerous unisexual flowers. The flower itself is not visible outwardly, as it blooms inside the infructescence. Although commonly referred to as a fruit, the fig is actually the infructescence or scion of the tree, known as a false fruit or multiple fruit, in which the flowers and seeds are borne. It is a hollow-ended stem containing many flowers. The small orifice (ostiole) visible on the middle of the fruit is a narrow passage, which allows the specialized fig wasp *Blastophaga psenes* to enter the fruit and pollinate the flower, whereafter the fruit grows seeds.

Fig is a tree of small dimensions, 10 to 30 ft (3-9 m) high, with numerous spreading branches and a trunk rarely more than 7 in (17.5 cm) in diameter. It contains copious milky latex. The root system is typically shallow and spreading, sometimes covering 50 ft (15 m) of ground, but in permeable soil some of the roots may descend to 20 ft (6 m). The deciduous leaves are palmate, deeply divided into 3 to 7 main lobes, these more shallowly lobed and irregularly toothed on the margins. The blade is up to 10 inch (25 cm) in length and width, fairly thick, rough on the upper surface, softly hairy on the underside. What is commonly accepted as a "fruit" is technically a synconium, that is, a fleshy, hollow receptacle with a small opening at the apex partly closed by small scales. It may be obovoid, turbinate, or pear-shaped, 1 to 4 inch (2.5-10 cm) long, and varies in color from yellowish-green to coppery, bronze, or dark-purple. Tiny flowers are massed on the inside wall. In the case of the common fig discussed here, the flowers are all female and need no pollination. There are 3 other types, the "Caprifig" which has male and female flowers requiring visits by a tiny wasp, *Blastophaga grossorum*; the "Smyrna" fig, needing cross-pollination by Caprifigs in order to develop normally; and the "San Pedro" fig which is intermediate, its first crop independent like the common fig, its second crop dependent on pollination. The skin of the fig is thin and tender, the fleshy wall is whitish, pale-yellow, or amber, or more or less pink, rose, red or purple; juicy and sweet when ripe, gummy with latex when unripe. Seeds may be large, medium, small or minute and range in number from 30 to 1,600 per fruit.

The edible fruit consists of the mature syconium containing numerous one seeded fruits (druplets). The fruit is 3–5 centimetres (1.2–2.0 inch) long, with a green skin, sometimes ripening towards purple or brown. *Ficus carica* has milky sap (laticifer). The sap of the fig's green parts is an irritant to human skin.

Origin and Distribution

The fig is believed to be indigenous to Western Asia and to have been distributed by man throughout the Mediterranean area. It has been cultivated for thousands of years, remnants of figs have been found in excavations of Neolithic sites traced to at least 5,000 B.C. As time went on, the fig-growing territory stretched from Afghanistan to southern Germany and the Canary Islands. Pliny was aware of 29 types. Figs were introduced into England some time between 1525 and 1548. It is not clear when the common fig entered China but by 1550 it was reliably reported to be in Chinese gardens. European types were taken to China, Japan, India, South Africa and Australia.

First figs in the New World were planted in Mexico in 1560. Figs were introduced into California when the San Diego Mission was established in 1769. Later, many special varieties were received from Europe and the eastern United States where the fig reached Virginia in 1669. The Smyrna fig was brought to California in 1881-82 but it was not until 1900 that the wasp was introduced to serve as the pollinating agent and make commercial fig culture possible. From Virginia, fig culture spread to the Carolinas, Georgia, Florida, Alabama, Mississippi, Louisiana and Texas. The tree was planted in Bermuda in early times and was common around Bahamian plantations in Colonial days. It became a familiar dooryard plant in the West Indies, and at medium and low altitudes in Central America and northern South America. There are fair-sized plantations on mountain sides of Honduras and at low elevations on the Pacific side of Costa Rica. From Florida to northern South America and in India only the common fig is grown. Chile and Argentina grow the types suited to cooler zones.

In Venezuela, fig is one of the fruits in the greatest demand by fruit processors. Because of the inadequate supply, a program was launched in 1960 to encourage commercial plantings. In 1976, fresh figs were regarded as highly desirable luxuries and were selling for $6.35 to $7.25 per lb ($14-$16/kg) in Colombia. The Instituto Colombiano Agropecuario had realized some years earlier that fig growing should be encouraged and had established an experimental plantation in 1973. The results were so favorable that they circulated an advisory bulletin to farmers in 1977, including improved methods of cultivation, costs of production and potential revenue.

Habitat

Common fig tree has been cultivated since ancient times and grows wild in dry and sunny areas, with deep and fresh soil; also in rocky areas, from sea level to 1,700 meters. It prefers light and medium soils, requires well drained soil, and can grow in nutritionally poor soil. Like all fig trees, *Ficus carica* requires wasp pollination of a particular species of wasp (*Blastophaga psenes*) to produce seeds. The plant can tolerate seasonal drought, and the Middle Eastern and Mediterranean climate is especially suitable for the plant. Situated in a favorable habitat, old specimens when mature can reach a considerable size and form a large dense shade tree. Its aggressive root system precludes its use in many urban areas of cities, but in nature helps the plant to take root in the most inhospitable areas. Common fig tree is mostly a phreatophyte that lives in areas with standing or running water, grows well in the valleys of the rivers and ravines having no water, having strong need of water that is extracted from the ground. The deep-rooted plant searches groundwater, in aquifers, ravines, or cracks in the rocks. The fig tree, with the water, cools the environment in hot places, creating a fresh and pleasant habitat for many animals that take shelter in its shade in the times of intense heat.

Ecology

Ficus carica is dispersed by birds and mammals that scatter their seeds in droppings. Fig fruit is an important food source for much of the fauna in some areas, and the tree owes its expansion to those that feed on its fruit. The common fig tree also sprouts from the root and stolon issues.

The infructescence is pollinated by a symbiosis with a kind of fig wasp (*Blastophaga psenes*). The fertilized female wasp enters the fig through the scion, which is a tiny hole in the crown (the ostiole). She crawls on the inflorescence inside the fig and pollinates some of the female flowers. She lays her eggs inside some of the flowers and dies. After weeks of development in their galls, the male wasps emerge before females through holes they produce by chewing the galls. The male wasps then fertilize the females by depositing semen in the hole in the gall. The males later return to the females and enlarge the holes to facilitate the females to emerge. Then some males enlarge holes in the scion, which enables females to disperse after collecting pollen from the developed male flowers. Females have a short time (<48 hours) to find another fig tree with receptive scions to spread the pollen, assist the tree in reproduction and lay their own eggs to start a new cycle.

Varieties

There are many cultivated varieties in each class of figs. In fact, over 700 varietal names are in use but many are synonyms. Here we need only present those that are suited to warm areas and do not require pollination. Most popular among these are 'Celeste' and 'Brown Turkey', followed by 'Brunswick' and 'Marseilles', described as follows:

'Celeste'

Pear-shaped, ribbed, sometimes with a short neck and slender stalk to 3/4 in (2 cm) long; the eye (opening at apex) is closed; the fruit is small to medium; the skin purplish-brown or bronze tinged with purple and covered with bloom; the pulp whitish or pinkish amber, of rich flavor and good quality; almost seedless. Main crop is heavy but of short duration. There is rarely an early, "breba", crop.

'Brown Turkey'

Road-pyriform, usually without neck; medium to large; copper-colored; pulp is whitish shading to pink or light red; of good to very good quality; with few seeds.

The tree is prolific. The main crop, beginning in mid-July, is large; the early, breba, crop is small. This cultivar is well adapted to warm climates. It is grown on all the islands of Hawaii.

'Brunswick' ('Magnolia')

Leaves narrow-lobed; fruits of main crop are oblique-turbinate, mostly without neck; fruit stalk thick, often swollen; fruit of medium size; bronze or purple-brown; pulp whitish near skin, shading to pink or amber; hollow in center; of fair to good quality; nearly seedless. Ripens over a long season. Breba crop poor; large, bronze-skinned; flesh light-red; coarse.

'Marseilles' ('White Marseilles', or 'Lemon')

Fruits of main crop round to oblate without neck; on slender stalks to 1/4 in (6 mm) long; of medium size. Those of breba crop, turbinate with short, thick neck and short stalk; yellow-green with small green flecks; pulp white, sweet; seeds large, conspicuous. Of fair quality.

In Queensland, 'Brown Turkey', 'Adriatic', 'Genoa' and 'Purple Genoa' perform very well.

'Adriatic' ('White Adriatic', or 'Grosse Verte')

Turbinate with short, thick neck and short stalk; above medium size; green to yellowish-green with red pulp; of distinctive flavor and very good quality. In early, minor, breba crop the fruits are oblique-pyriform, large, green, often tinged with purplish-red with dark-red pulp and strong flavor.

'Genoa' ('White Genoa')

Pyriform or turbinate, very faintly ribbed; neck thick and short, or absent; above medium in size; skin downy, greenish-yellow; pulp greenish-white near skin, mostly amber tinged with red; hollow; of fair quality. Fruits of breba crop oblique-obovate with thick neck and short stalk; yellowish-green externally; pulp light-red; of fair to good quality.

'Purple Genca' ('Black Genoa'; 'Black Spanish')

Oblong, broad at apex, narrow at base; large; very dark-purple with thick blue bloom; pulp yellowish becoming reddish to red at the center; juicy, with sweet, rich flavor.

At Saharanpur, India, 'Brown Turkey', 'Bengaluru', 'Black Ischia' and 'Lucknow' are successfully grown. Around Bombay, there is only one variety, 'Poona'.

'Black Ischia' ('Blue Ischia')

An Italian variety; main crop is elongated pear shaped with many noticeable ribs; short neck and short to medium stalk; large, 2½ in (6.35 cm) long and 1½ in (3.8 cm) wide; dark purple-black except at the apex where it is lighter and greenish; there are many golden flecks; skin is wholly coated with thin, dark-blue bloom; eye open, with red-violet scales; pulp is violet-red, of good quality. In the breba crop, there are few ribs and mostly indistinct; the fruit is small, about 1½ in (3.8 cm) long and of the same width at the apex; the pulp is red to greenish-amber; of poor flavor. The tree is particularly ornamental and the leaves are glossy, only shallowly 3 lobed. A heavy bearer.

'Poona'

Bell-shaped, of medium size, weighing about 1½ oz (42 g); thin-skinned; light-purple with red flesh, of sweet, good flavor.

Recommended Varieties

- ✰ 'Brown Turkey', which produces abundant medium to large fruits. This type works best in warm climates.
- ✰ 'King', which is well adapted for the cooler conditions found in the northwest. It produces medium fruit that are sweet and rich in flavor.
- ✰ 'Kadota', which vigorously produces small to medium fruit. Its figs are rich and sweet and it is the most commonly canned fig.

There are basically three types of edible figs:

- ✰ Persistent (or common) figs have all female flowers that do not need pollination; the fruit develop through parthenocarpic means. This is a popular horticulture fig for home gardeners. Adriatic, Black Mission, Brown Turkey, Brunswick, and Celeste are some representative cultivars
- ✰ Caducous (or Smyrna) figs require crosspollination by the fig wasp with pollen from caprifigs for the fruit to mature. Some cultivars are Calimyrna (the Great Central Valley Smyrna fig), Marabout, and Zidi
- ✰ Intermediate (or San Pedro) figs set an unpollinated breba crop, but need pollination, at least in most regions, for the later main crop. Examples are Lampeira, King, and San Pedro
- ✰ 'Alma'
- ✰ 'Brown Turkey'
- ✰ 'Celeste'
- ✰ 'Desert King'
- ✰ 'Italian Black'
- ✰ 'Italian White'
- ✰ 'Kadota': used in fig rolls, dries well
- ✰ 'Lemon Fig': also known as 'Blanch' or 'Marseilles'
- ✰ 'Mission': black, sweet, commonly dried
- ✰ 'Timla in Kumaon'

Climate

In southern India, 'Marseilles' flourishes on hills above 5,000 ft. (1,525 m). In tropical areas generally, figs thrive between 2,600 and 5,900 ft. (800-1,800 m). The tree can tolerate 10° to 20° of frost in favorable sites. It should have a dry climate with light early spring rains if it is intended for the production of fresh fruit. Rains during fruit development and ripening are detrimental to the crop, causing the fruits to split. The semi arid tropical and subtropical regions of the world are ideal for fig growing if means of irrigation are available. But very hot, dry spells will cause fruit-drop even if the trees are irrigated.

Soils

The fig can be grown on a wide range of soils; light sand, rich loam, heavy clay or limestone, providing there is sufficient depth and good drainage. Sandy soil that is medium dry and contains a good deal of lime is preferred when the crop is intended for drying. Highly acid soils are unsuitable. The pH should be between 6.0 and 6.5. The tree is fairly tolerant of moderate salinity.

Propagation

Fig plants are easy to propagate through several methods. Propagation using seeds is not the preferred method since vegetative methods exist that are quicker and more reliable, that is, they do not yield the inedible caprifigs. However, those desiring to can plant seeds of dried figs with moist sphagnum moss or other media in a zip lock bag and expect germination in a few weeks to several months. The tiny plants can be transplanted out little by little once the leaves open, and despite the tiny initial size can grow to 1 foot (30 cm) or more within one year from planting seeds.

For propagation in the mid summer months, air layer new growth in August (mid-summer) or insert hardened off 15–25 cm (6-10 inches) shoots into moist perlite or a sandy soil mix, keeping the cuttings shaded until new growth begins; then gradually move them into full sun. An alternative propagation method is bending over a taller branch, scratching the bark to reveal the green inner bark, then pinning the scratched area tightly to the ground. Within a few weeks, roots will develop and the branch can be clipped from the mother plant and transplanted where desired.

For spring propagation, before the tree starts growth, cut 15–25 cm (6-10 inches) shoots that have healthy buds at their ends, and set into a moist perlite and/or sandy soil mix located in the shade. Once the cuttings start to produce leaves, bury them up to the bottom leaf to give the plant a good start in the desired location.

Fig trees have been raised from seed, even seed extracted from commercial dried fruits. Ground or air-layering can be done satisfactorily, and rapid mass multiplication by tissue culture has been achieved in Greece, but the tree is commonly propagated by cuttings of mature wood 2 to 3 years of age, 1/2 to 3/4 in (1.25-2 cm) thick and 8 to 12 in (20-30 cm) long. Planting must be done within 24 hours but, first, the upper, slanting end of the cutting should be treated with a sealant to protect it from disease, and the lower, flat, end with a root-promoting hormone. Trees of unsatisfactory varieties can be top worked by shield- or patch-budding, or cleft or bark-grafting.

Culture

Cuttings are raised in nursery beds and are set out in the field after 12 or 15 months. They may be spaced from 6 ft. to 25 ft. (1.8-7.5 m) apart depending on the cultivar and the fertility of the soil. A spacing of 13 ft. x 13 ft. (4 m x 4 m) allows 625 trees ha^{-1}. In Colombia, the trees are at 10 ft. x 10 ft. (3 x 3 m) on level land, 10 ft. x 13 ft. (3x4 m) on slopes. Fruiting will commence in less than a year from planting out. Young plants will benefit from shading with palm fronds or other material until they are well established. A fertilizer dose of 60 g of 10:30:10 or 10:20:20 $N:P_2O_5:K_2O$ is recommended for young plants and 100 g for adults, micronutrients at the rate of 30 g per tree every 6 months.

Fig trees are cut back severely in fall or winter, depending on whether the crop is desired the following summer or fall. Branches are often notched to induce lateral branching and increase the yield. If there are heavy rains, drainage ditches should be dug to prevent water-logging. Fig trees remain productive up to 12 or 15 years of age and thereafter the crop declines though the trees may live to a very advanced age.

Season

Fig trees usually bear 2 crops a year, the early season ("breba") fruits being inferior and frequently too acid, and only those of the second, or main, crop of actual value.

In Colombia and Venezuela, some fruits are borne throughout the year but there are 2 principal crops, one in May and June and the other in December and January.

Large-scale fig producers in California spray ethephon to speed up ripening and then wind-machines are drawn past the trees or helicopter overflights are made to hasten fruit drop, thus, shortening the harvest period by as much as 10 days in order to avoid impending rain and insect attack. Proper timing of the growth regulator is crucial to fruit quality.

Training Pruning

Figs require little pruning just enough to keep growth within bounds, to keep the crown open to sunlight and air, and to remove dead wood. Pruning should be done during the winter months, preferably after the coldest weather is past but before growth starts in the spring.

Most fig cultivars bear two crops a year: a light one in early summer borne on previous season's wood and the second crop in mid or late summer borne on current season's wood.

Though fig plants can be trained to either a tree or a bush form, the tree form is not practical for the piedmont. In this region, fig plants are occasionally frozen back, making the tree form difficult to maintain.

Begin training to bush form at time of planting by cutting off one-third of the young plant. This forces shoots to grow from the base of the plant. Let these shoots grow through the first season. Then, during the late winter after the first growing season, select three to eight vigorous, widely spaced shoots to serve as leaders. Remove all other shoots.

Be sure the leaders are far enough apart to grow to 3 to 4 inches in diameter without crowding each other. If they are too close together, the leaders cannot grow thick enough to support themselves and their crop and tend to fall over or split off under stress of high winds. If this happens, remove the damaged leader and select a new one late the next winter by choosing one of the many suckers that arise annually.

Beginning the second year after planting, if more branching is desired, head back the bush each spring after danger of frost is past but before growth has started. Do this by removing about one-third to one-half the length of the annual growth. Also, prune out all dead wood, and remove branches that interfere with growth of the leaders. Cut off low-growing lateral branches and all sucker growth that is not needed for replacement of broken leaders. Do not leave bare, unproductive stubs when you prune. These stubs are entry points for wood decay organisms. Make all pruning cuts back to a bud or branch.

Harvesting and Yield

The fruits may be picked from the tree or gathered normally or by mechanical sweepers after they fall to the ground. 'Brunswick' is so tender it must be picked when slightly unripe in order to be firm enough for processing. Workers must wear gloves and protective clothing because of the latex. Harvested fruits are spread out in the shade for a day so that the latex will dry a little. Then they are transported to processing plants in wooden boxes holding 22 to 33 lbs (10-15 kg). In India, a fig tree bears 180 to 360 fruits per year. Venezuelan growers expect 6-8 kg tree^{-1}.

Harvest/Storage

- ☆ You should harvest figs when they are fully ripe. The figs should be fully colored and slightly soft to the touch.
- ☆ When picking figs, wear gloves or long sleeves because the sap from the tree can irritate your skin.
- ☆ Figs are very perishable. Store figs in the refrigerator; they will keep for 2 to 3 days.
- ☆ For long-term storage, you can freeze figs whole for later use. Another storage method is to dry the figs. You can also can your own figs.

Keeping Quality

Fresh figs are very perishable. At 40° to 43°F (4.44°-6.11°C) and 75 per cent relative humidity, figs remain in good condition for 8 days but have a shelf life of only 1 to 2 days when removed from storage. At 50°F (10°C) and relative humidity of 85 per cent, figs can be kept no longer than 21 days. They remain in good condition for 30 days when stored at 32° to 35° F (0°-1.67° C). If frozen whole, they can be maintained for several months.

Pests and Diseases

Root-knot Nematode

Fig trees are prone to attack by nematodes (especially *Meloidogyne* spp.) and, in the tropics, have been traditionally planted close to a wall or building so that the roots can go underneath and escape damage. Heavy mulch will serve equally well. Today, control is possible with proper application of nematicides.

Root-knot nematodes are the primary pest of fig trees in the sandhills and coastal plain. An on the spot diagnosis of root-knot infection is possible. Dig up a few roots and look for the characteristic galling or swelling caused by the nematode. There is no other similar problem in figs.

Infected fig trees cannot be cured with chemical treatment. Pruning back the top to balance it with the weakened root system and attentive watering and fertilization may prolong the life of a root knot infected fig tree. Usually, however, they will die sooner or later regardless of the care they receive.

When planting a new fig tree, select a site as far as possible from any old garden sites. Take a nematode sample in this site. If root knot nematodes are present, do not plant figs.

In India, a stem-borer, *Batocera rufomaculata,* feeds on the branches and may kill the tree. Lepidopterous pests in Venezuela include the fig borer, *Azochis gripusalis,* the larvae of which feed on the new growth, tunnel down through the trees to the roots and kill the tree. Another, called *cachudo de la higuera,* has prominently horned larvae up to 3 $^{1}/_{8}$ in (8 cm) long that can destroy a fig tree in a few days. There are also coleopterous insects of the genera *Epitrix* and *Colaspis* that perforate and severely damage the leaves and shoots. Scale insects include *Asterolecanium* sp. which attacks the bark of trees weakened by excessive humidity or prolonged drought, and the lesser enemy, *Saissetia haemispherica.*

A common and widespread problem is leaf rust caused by *Cerotelium fici;* bringing about premature leaf fall and reducing yields. It is most prevalent in rainy seasons. Leaf spot results from infection by *Cylindrocladium scoparium* or *Cercospora fici.* Fig mosaic is caused by a virus and is incurable. Affected trees must be destroyed.

The dried fruit beetle, or sour bug, *Carpophilus* spp., enters the fruit through the eye and leads to souring and smut caused by *Aspergillus niger*. This fungus may attack ripening fruits.

Food Uses

Some people peel the skin back from the stem end to expose the flesh for eating out of hand. The more fastidious eater holds the fruit by the stem end, cuts the fruit into quarters from the apex, spreads the sections apart and lifts the flesh from the skin with a knife blade, discarding the stem and skin. Commercially, figs are peeled by immersion for 1 minute in boiling lye water or a boiling solution of sodium bicarbonate. In warm, humid climates, figs are generally eaten fresh and raw without peeling, and they are often served with cream and sugar. Peeled or unpeeled, the fruits may be merely stewed or cooked in various ways, as in pies,

Table 8.1: Diseases, Nature of Damage and Control Measures of Fig

Sl.No.	Name of Diseases/Pest	Nature of Damage	Control Measures
1.	**Fig Rust**	This fungus attacks the leaves, usually in late summer. Severely infected leaves turn yellow brown and drop. The underside of the fallen leaves will have numerous small, somewhat raised, reddish brown spots. These spots are often covered with a dusty golden yellow mass of rust spores. Fig rust are usually not fatal, but repeated epidemics will weaken the plant. In any given year, heavy leaf drop from rust will reduce size and quality of the fruit. Unless fig rust is an annual problem, spraying is not warranted.	– Spraying of clorothodonil 2 g lit^{-1} or Bavistin 1 g lit^{-1} or Bavisitin water – Bordeux spray (5 : 5 : 50). (Copper sulphate : Lime : Water)
2.	**Fig Fruit Souring**	This is caused by yeasts that are spread by insects. Souring becomes noticeable as the figs begin to ripen. A souring fig will often show gas bubbles and/or scummy masses oozing from the eye. These figs will give off an offensive fermented odour.	Souring cannot be controlled with chemical sprays. The only control is to grow fig cultivars that have a tight or closed eye that prevents insects from entering the fig fruit.
3.	**Pink Blight**	This is a fungus that appears as a dirty-white to pale pink velvety growth on dying and dead twigs. It usually occurs in the interior of the tree.	Prune out infected branches and prune the tree to allow good air movement within the tree. Spraying of copper oxichloride 2.5 g lit^{-1} or Manocozeb 2.5 g lit^{-1}
4.	**Leaf Blight (Thread Blight)**	This is a fungal disease that attacks leaves and fruit. Infection may start as a semicircular brown spot at the base of the leaf. Some leaves shrivel and die while others may be covered with brown spots that break out to leave irregular holes. During hot, wet weather, leaves can die and drop very quickly. Dead leaves are often matted together and held to the tree by thread like strands similar to spider webs.	Spraying of Dithane M-45 3 g lit^{-1} or Bavistin 2 g lit of water.

puddings, cakes, bread or other bakery products, or added to ice cream mix. Home owners preserve the whole fruits in sugar syrup or prepare them as jam, marmalade, or paste. Fig paste (with added wheat and corn flour, whey, syrup, oils and other ingredients) forms the filling for the well known bakery product, "Fig Newton". The fruits are sometimes candied whole commercially. In Europe; western Asia, northern Africa and California, commercial canning and drying of figs are industries of great importance.

Some drying is done in Poona, India, and there is currently interest in solar-drying in Guatemala. Usually, the fruits are allowed to fully ripen and partially dehydrate on the tree, then are exposed to sulphur fumes for about a half hour, placed out in the sun and turned daily to achieve uniform drying, and pressed flat during the 5- to 7-day process. 'Black Mission' and 'Kadota' figs are suitable for freezing whole in syrup, or sliced and layered with sugar.

Dried cull figs have been roasted and ground as a coffee substitute. In mediterranean countries, low-grade figs are converted into alcohol. An alcoholic extract of dried figs has been used as a flavoring for liqueurs and tobacco.

Toxicity

The latex of the unripe fruits and of any part of the tree may be severely irritating to the skin if not removed promptly. It is an occupational hazard not only to fig harvesters and packers but also to workers in food industries, and to those who employ the latex to treat skin diseases.

Other Uses

Seed Oil

Dried seeds contain 30 per cent of a fixed oil containing the fatty acids: oleic, 18.99 per cent; linoleic, 33.72 per cent; linolenic, 32.95 per cent; palmitic, 5.23 per cent; stearic, 2.18 per cent; arachidic, 1.05 per cent. It is an edible oil and can be used as a lubricant.

Leaves

Fig leaves are used for fodder in India. They are plucked after the fruit harvest. Analyses show: moisture, 67.6 per cent; protein, 4.3 per cent; fat, 1.7 per cent; crude fiber, 4.7 per cent; ash, 5.3 per cent; N-free extract, 16.4 per cent; pentosans, 3.6 per cent; carotene on a dry weight basis, 0.002 per cent. Also present are bergaptene, stigmasterol, sitosterol, and tyrosine.

In southern France, there is some use of fig leaves as a source of perfume material called "fig-leaf absolute"—a dark-green to brownish-green, semi-solid mass or thick liquid of herbaceous-woody-mossy odour, employed in creating woodland scents.

Latex

The latex contains caoutchouc (2.4 per cent), resin, albumin, cerin, sugar and malic acid, rennin, proteolytic enzymes, diastase, esterase, lipase, catalase, and peroxidase. It is collected at its peak of activity in early morning, dried and powdered for use in coagulating milk to make cheese and junket. From it can be isolated the protein-digesting enzyme *ficin* which is used for tenderizing meat, rendering fat, and clarifying beverages.

In tropical America, the latex is often used for washing dishes, pots and pans. It was an ingredient in some of the early commercial detergents for household use but was abandoned after many reports of irritated or inflamed hands in housewives.

Medicinal Uses

The latex is widely applied on warts, skin ulcers and sores, and taken as a purgative and vermifuge, but with considerable risk. In Latin America, figs are much employed as folk remedies. A decoction of the fruits is gargled to relieve sore throat; figs boiled in milk are repeatedly packed against swollen gums; the fruits are much used as poultices on tumours and other abnormal growths. The leaf decoction is taken as a remedy for diabetes and calcifications in the kidneys and liver. Fresh and dried figs have long been appreciated for their laxative action.

Table 8.2: Food Value Per 100 g of Edible Portion

	Fresh	*Dried*
Calories	80	274
Moisture	77.5-86.8 g	23.0 g
Protein	1.2-1.3 g	4.3 g
Fat	0.14-0.30 g	1.3 g
Carbohydrates	17.1-20.3 g	69.1 g
Fiber	1.2-2.2 g	5.6 g
Ash	0.48 0.85 g	2.3 g
Calcium	35-78.2 mg	126 mg
Phosphorus	22-32.9 mg	77 mg
Iron	0.6-4.09 mg	3.0 mg
Sodium	2.0 mg	34 mg
Potassium	194 mg	640 mg
Carotene	0.013-0.195 mg	—
Vitamin A	20-270 I.U.	80 I.U.
Thiamine	0.034-0.06 mg	0.10 mg
Riboflavin	0.053-0.079 mg	0.10 mg
Niacin	0.32-0.412 mg	0.7 mg
Ascorbic Acid	12.2-17.6 mg	0 mg
Citric Acid	0.10-0.44 mg	

Note: There are small amounts of malic, boric and oxalic acids.

References

Armstrong, W.P., 2006. Sex determination and life cycle of *Ficus carica*. Aug., 2006. <http://waynesword.palomar.edu/pljun99b.htm>.

Bob Polomski, School of Agricultural, Forestry, and Environmental Science, Clemson University. 10/12. HGIC 1353.

Cobanoglu, F.,B.Sahin., H. Kocatas, M. Ozen., 2004. Productivity and Quality Parameters on Fig Young Plant in Tube. Gazi Osmanpasa University, *Journal of Agricultural Faculty*, 21(1): 1-8.

Fruit and Nut Research and Information Center. Fig Fact Sheet. http://fruitandnuts.ucdavis.edu/fig3.html.

Gawade, M.H., and D.P. Waskar. 2002. Physico-chemical characteristics of indigenous and exotic varieties of fig grown under conditions. *South Indian Hort.* 50: 505-508.

Gawade, M.H., and D.P. Waskar, 2004. Dried fig (*Ficus carico* L.) storage under various environments. *J. Food Sci. Technol. Mysore* 41: 426-429.

Gawade, M.H., and D.F. Waskar, 2005. Effect of different varieties and pretreatments on yield and quality of dried fig fruits. *Indian J.Agr. Res.* 39: 138-141.

Genna, A., A. Maestrelli, P. De Vecchi, and M. Bruno, 2005. Quality of "Cosenza dried fig". A sensory añd physicochemical approach, p. 72. In: *Third Int. Syinp. Fig.* 16-20 May, Vilamouira,- Portugal.

Jigna P., Rathish N., Sumitra, C., 2005. Preliminary screening of some folklore medicinal plants from western India for potential antimicrobial activity. *Indian J Pharmacol*, 37(6): 408-409.

Kislev, Mordechai E.; Hartmann, Anat and Bar-Yosef, Ofer, 2006a. Early Domesticated Fig in the Jordan Valley. *Science* **312**(5778): 1372. doi:10.1126/science.1125910 PMID 16741119 (HTML abstract) Supporting Online Material.

Kislev, Mordechai E.; Hartmann, Anat and Bar-Yosef, Ofer, 2006b. Response to Comment on "Early Domesticated Fig in the Jordan Valley". *Science*, **314**(5806): 1683b. doi:10.1126/science.1133748 PDF fulltext.

Lev-Yadun, Simcha; Ne'eman, Gidi; Abbo, Shahal and Flaishman, Moshe A. (2006): Comment on "Early Domesticated Fig in the Jordan Valley". *Science* **314**(5806): 1683a. doi:10.1126/science.1132636 PDF fulltext.

Morton, J. 1987. Fig. p. 47–50. In: Fruits of warm climates. Julia F. Morton, Miami, FL.

Tous, J. and L. Ferguson. 1996. Mediterranean Fruit p. 416 – 430. In: J. Janick (ed.), Progress in New Crops. ASHS Press, Arlington, VA.

Vinson, Joe A. (1999): Functional food properties of figs. *Cereal Foods World* **44**(2): 82-87. PDF fulltext

Vinson, Joe A.; Zubik, Ligia; Bose, Pratima; Samman, Najwa and Proch, John (2005): Dried fruits: excellent *in vitro* and *in vivo* antioxidants. *J. Am. Coll. Nutr.* **24**(1): 44–50. PMID 15670984 PDF fulltext.

Fig Orchards

Fig Fruits

9
Karonda

A.B. Bhosale, R.A. Durgude and S.D. Kale

Department of Horticulture,
MPKV, Rahuri, Maharashtra

Carissa Carandas Linn. (Karaunda) is a widely used medicinal plant by tribals throughout India and popular in various indigenous system of medicine like Ayurveda, Unani and Homeopathy. All parts of the plant are used in traditional medicine. Traditionally, the plant has been used in the treatment of scabies, intestinal worms, pruritus, biliousness and also used as antiscorbutic, anthelmintic. The notable biological activities reported are analgesic, anti inflammatory, anti pyretic, cardiotonic and histamine releasing. The plant is also an alternative source of oil, hydrocarbon and phytochemicals. Medicinal plants are a major source of biodynamic compounds of therapeutic values and are basis of many traditional medicines throughout the world for thousands of years. Herbal medicine is still the mainstay of about 75–80 per cent of the world population, mainly in the developing countries for primary healthcare not because they are inexpensive but also for better cultural acceptability, better compatibility with the human body and minimal side effects. However, among the estimated 250,000-400,000 plants species, only 6 per cent have been studied for biological activity and about 15 per cent have been investigated phytochemically. Karonda (*Carissa carandas*) is found in throughout India mainly in the semi-arid regions. Karonda tree is widely cultivated in the home gardens, farmer's fields and orchards as hedge plant. Karonda is commonly used as a condiment or additive to Indian pickles and spices. It is a very hardy, drought tolerant plant that thrives well in a wide range of soils. Karonda is an important minor fruits. It's very nutritious, especially as a rich source of iron. It is very hardy tree and grows well in arid and semi-arid regions. It is an ideal windbreak plant especially for high dense orchards.

Karonda (*Carissa carandas*) is an evergreen spiny shrub or a small tree. It belongs to the family *Apocyanacae*. It is very popular as a protective hedge plant. When allowed to grow fully, it attains a height of 3 to 6 meter. It yields a heavy crop of berry like fruits which are edible and attractive in colour.

It is an indigenous fruit of India. It is widely grown in India, South Africa and Malaysia. In India, it is said to grow wild in Bihar, West Bengal and South India. It flowers from February to April and makes use of natural moisture availability period and therefore, highly suitable for growing in arid regions. Since its organized orcharding is not in vogue, data on its area covered in India are not available.

The karonda (*Carissa carandas*) is a non traditional fruit crop which thrives well as a rainfed crop. Once established, the plant hardly needs any care and gives yield with minimum management. This fruit crop is a monitory reward to tribal communities in various parts of India.

It is an indigenous, underutilized fruit of India. It is a hardy multipurpose horticultural bush flourishing well without much care in and semi-arid areas. At present, emphasis is being laid on protection and production value of karonda plant. It is grown as hedge for beautification which serves as one of the best bio-fencing materials because of thorny nature. It is also grown as an ornamental plant due to its beautiful cherry like fruits.

Composition and Uses

The composition of dried karonda fruit is given are as below:

Table 9.1: Composition of Dried Karonda Fruit

Sl.No.	*Components*	*Quality*
1.	Moisture (per cent)	18.2 per cent
2.	Carbohydrate (g)	2.3 per cent
3.	Protein (g)	9.6 per cent
4.	Fat (g)	2.8 per cent
5.	Fibre (g)	—
6.	Total mineral matter (g)	67.1 per cent
7.	Calcium (mg)	0.16 per cent
8.	Phosphorus (mg)	0.06 per cent
9.	Iron (mg)	39.1 per cent
10.	Calorific values/100mg	364

The composition and food value of ripe and dry karonda fruits are given in Table 9.2.

Uses

Ripe fruits are sub acidic to sweet in taste with peculiar aroma. The fruits may be eaten as a dessert when ripe or used in the preparation of fruit products such as jelly, sauce, carissa, cream or jellied salad. The unripe fruits can be used for pickles, sauces and *Chutney.* The dried fruits may become a substitute for raisins (Cheema and Cheema, 1971). The fruit can he candied just like cherry. Girdhari Lal *et al.* (1967) described the procedure for making karonda preserve. The wine prepared from ripe fruits contains about 14.5 to 15 per cent alcohol and is very much liked by wine fanciers (Nalawadi and Jayasheeln. 1975).

Table 9.2: Proximate Composition of Karonda Fruit

Constituents	*Values per 100 g of Edible Portion*		*Minerals*	*mg/100g*	
	Fresh Fruit	*Dry Fruit*		*Fresh Fruit*	*Dry Fruit*
Moisture	91.0	18.2	Calcium	21.0	0.16
Protein	1.1	2.3	Phosphorus	38.0	0.06
Carbohydrate	2.9	67.1	Iron	39.1	39.1
Fat	2.9	9.6			
Fibre	1.5				
Calorific Values (per 100 g)	42.0	364			

The unripe fruits yield milky white latex which can also be used in preparing chewing gum and rubber. Fruits can also be used in dyeing and tanning industries. Root extracts are used in lumbago, chest complaints and veneral diseases. Very useful to cure anaemia, its fruits have antiscorbutic properties also.

The wood of the karonda plant is white, hard and smooth. It is used for making spoons and combs.

Fruit

Karonda fruit is a rich source of iron and contains a fair amount of vitamin C, and therefore, very useful for curing of anaemia and has antiscorbutic properties. Enthomedically, fruits are used as an astringent, antisorbutic and as a remedy for biliousness. Mature fruit contains high amount of pectin and therefore, besides being used for making pickles and chutney, it can be exploited for making jelly. Raw fruit can be used for making chutney.

Leaf

Karonda leaves are palatable and contain a fair amount of nutrient which is browned by sheeps, goats and camels in arid and semi arid areas. If fruits fall on ground, enrich soil by adding organic carbon and outer mineral nutrients. It reduces evaporation from soil underneath due to dense foliage. Leaves are used for feeding silk worm. Leaf decoction is used against intermittent fever, diarrhoea, oral inflammation and ear ache.

Root

Roots are heavily branched which improves binding capacity of soil. During rainy season, it minimizes runoff and soil loss on sloppy land. Roots serve as stomachic, vermifuge, and as remedy for itches and insect repellent. Roots contain salicylic acid and cardiac glycosides causing a slight decrease in blood pressure. Report reveals carissone; the *D-glycoside* of *B-sitosterol;* glucosides of oderside H; presence of carindone; lupeol; ursolic acid, and its methyl ester; also carinol, a phenolic lingan in roots and karonda plants.

Wood

Wood is hard and used for making combs, toys, utensils and small agricultural implements. It is also used as a fuel wood in dry region.

Karonda fruits are a rich source of pectin, carbohydrates and vitamin C. They are sub-acidic and sweet when ripe. Dried karonda fruits are very rich source of iron. Fruits are also used for pickles, chutney, tart, pudding and jelly. The fruits can be dried and preserved with salt. Besides preserving, the fresh fruits can be cooked as vegetables and used in place of cherry for decoration of sweets and pastries. In this way the karonda is a multipurpose shrub which is useful as a protective and ornamental hedge and as orchard fruit with proper care, its production can be increased many fold.

Taxonomy and Morphology

Botanical name of karonda is *Carissa conjesta* Wight (Syn. *C. carandas* Auct., formerly known as *C. carandas* L.) It belongs to family *Apocynaceae* and has over 30 species originating in South Africa, Asia and Malaya. A few of the important cultivated species of karonda are *carissa grandiflora* DC, *C. bispinosa* Desf., *C. spindrum* DC, *C. ovata, C. Mulls* Vahl. *C. inermis* Vahl. Syn. *C. macroplzylla, C. paucinervia* D.C., *C. spinarum* L. Syn. C. *diffiisa.*

Karonda grows as a small shrub or tree attaining a height of 2-3 m. Botanically it is a large dichotomously branched evergreen shrub with short stem and strong thorns in pairs simple or forked up to 5 cm long in pair in the axil of leaves. Leaves are simple, evergreen, oval or elliptic in shape, 1 to 3 cm long, dark green, leathery, glossy on upper surface, lighter green and dull on under surface. New emerging leaves are slightly reddish in colour. Cohen and Arzee (1980) reported that in *C. grandiflora,* individual branches of sympodial shoot usually bear 2-3 pairs of decussate leathery ovate foliage leaves and a pair of inconspicuous ovate leaves from whose dzits, the branched thorny shoots grow.

In Carissa, flowers are small (3-5 cm in diameter), found in corymbose cymes are pedicellate, bracteate, hermaphrodite, complete, hypogynous and pentamerous. In *C. grandiflora,* flowers could be borne terminally between thorny branches or forked thorn or on branches in place of a thorny tip. Flowering in C. *carandas* takes place in spring *C. bispinosa* also flowers during February-March under Allahabad conditions). Flower bud takes 19-21 days to develop to full bloom stage in *C. conjesta* (Syn. *C. carandas*) whereas in *C. edulis,* it takes about 19 days In *C. congesta,* flowers open mostly between 8:10 pm (Srivastava, 1966), whereas in *C. edulis,* anthesis time was reported between 12:30–4:30 pm under Jodhpur condition and it took 3-4 hours for complete bud opening. Only 36 per cent bud transformed into fruits and remaining dropped off at different stages of development.

Origin and Distribution

Karonda grows well under tropical as well as sub-tropical conditions. It is widely grown in South Africa, Australia, India, Malaysia, Sri Lanka and Myanmar (Burma). It is native to India and grows wild in Maharashtra, Rajasthan. Uttar Pradesh, Bihar and South India. It is preferred very much as a protective hedge in Gujarat and Punjab. Regular plantations of Karonda are very common in Varanasi district of Uttar Pradesh. Karonda is reported to flourish well in the Western Ghats and commonly found in hills, roadside and riverside

Origin of karonda (C. *carandas* L.) is India, Natal plumv (C. *grandiflora*) came to India from South Africa and C. *edulis* Vahl. was introduced from the USA (Pareek and Sharma, 1993). Its occurrence is also common in Burma, Malacca and dry areas of Sri Lanka. It is grown in Philippines, Thailand, Cambodia, South Vietnam and in East Africa. In India, it is wild in Bihar, South Eastern part of Chattisgarh, West Bengal and South India. It is also grown on a limited scale in Rajasthan, Gujarat, Punjab, Uttar Pradesh and in lower hills of Shiwalik. It is a good alternative for resource poor farmers in saline, sodic soils and degraded uplands.

Area and Production

Information on area and production are not available. In India, karonda is a subsistence crop meeting local demands. Although, it is grown in different parts of the country but due to lack of systematic plantation, no official estimates are available. It is also difficult to make a realistic estimate of the area under this crop due to scattered nature of cultivation.

Soil and Climate

Karonda thrives well in a wide range of soils. Deep, fertile and well soils are ideal for its cultivation. It can also grow successfully on poor, sandy and rocky soils. In Florida, the plant grows vigorously on sand or limestone. It is classified according to salt tolerance limit, karonda as medium salt tolerant crop which could be grown upto 30-40 Exchangeable sodium per cent (ESP) and Electrical conductivity (EC) 9-12 dSm^{-1}. C *grandiflora* is affected little by salinity of 8 dSm^{-1}. There is direct relationship between transpiration and ESP levels as reported by Valia *et al.* (1993). Crop is well adapted to varying climatic conditions from an altitude of 300 to 1800 m in Himalayas. Crop is susceptible to frost if prolonged for long period. Karonda is classified as dryland crop for scarcity zone by Tambe *et al.* (1993). Plants possess drought avoidance organs like thorn on stem and branch and less stomata on lower surface of leaf which allow plants to perform better under drought conditions of Rajasthan and Gujarat.

It can be grown on a wide range of soils including poor, rocky, sandy soils of arid regions. However, on deep loamy or alluvial soils, its growth is better and fruit yield is higher. It is very well suited to waste lands and on field boundaries. It can tolerate saline sodic soils (pH upto 8.5) to a greater extent. Since karonda is very hardy and drought tolerant. It thrives well throughout the tropical and subtropical climates.

Owing to the xerophytic nature, plants of karonda are well adopted to the arid and semi-arid climates. Heavy rainfall and waterlogged conditions are not desirable. Being a hardy plant. It can thrive well under drought conditions.

Karonda is a very hardy shrub and can be successfully grown in tropical and subtropical climates. However, heavy rainfall and waterlogged conditions are harmful. Soil is not a limiting factor for its cultivation. It can be grown in any soil except very alkaline or sodic soils. But for better production, fertile sandy loam soil is better.

Species and Varieties

A. Species

There are some 20 species of karonda, of which 5 are of Indian origin including *Carissa carandas*. Some important cultivated species are described here.

1. *Carissa carandas*

Fruits are grape green in colour when young. White and pink when approaching maturity and nearly black when ripe. The fruits ripen in July-August.

2. *Carissa grandiflora*

The fruits are dark red with papery skin and a few small seeds. Fruits are available throughout the year.

3. *Carissa edulis*

Flowers are white, pink and scented. Flowers are borne in auxiliary clusters of 10 to 25. Fruits are oval in shape. They are red in colour but turn black at maturity. It is a straggling shrub.

4. *Carissa ibispinosa*

The flowers are small and fruits are borne in clusters. The fruits resemble that of *C. grandiflora.*

5. *Carissa ovata*

Fruits are small and are used for jam making. The origin of this species is said to be Australia.

6. *Carissa spindrum*

This species bears small fruits. It is native of India. This is the hardiest species of Carissa and is known to thrive in poor and rocky soil.

The genus *Carissa* to which karonda belongs is a member of the family *Apocynaccae*. The genus has nearly 30 species originating is South Africa, Australia tropical Asia and Malaysia. The five species of this genus of Indian origin including *C. carandas. C'. inermis. C'. paucinerva, C'. spinarum* grow wild in India; the first two Western peninsula and third in the dry tracts (Singh and Arera, 1978). *C. grandiflora* (Natal plum) came to India from South Africa. C. *edulis* has been introduced from USA. Different species of *carissa* are grown for their small berry like.

B. Varieties

Although there is no known variety in karonda, but based on the colour of the fruits, three types are commonly grown. These are (1) green fruited, (2) white'fruited and (3) pink fruited. The fruits of these strains are similar in size but there is a difference in the production of various plants of these promising clones.

There are no well established varieties of karonda. Cultivated types are classified on the basis of fruit colour - green fruited, whitish fruits with pink blush and dark purple fruited. Maroon (Narendra Selection) and Nos. 13, 16, 12 and 2 have been identified as promising types at MPKV, Rahuri (Maharashtra). At Dr. Balasaheb Savant Konkan Krishi Vidvapeeth (Dr.BSKKV), Dapoli, three plants namely No.2, No.5 and No.2 (from another location) were reported to be promising by Bhagwat (1984). Similarly, Joshi *et al.* (*1986*) reported that ripe fruits of three genotypes i.e., K-1, K-2, K-3 were highly palatable. Sawant (1998) reported genotype No.21, 25, 46, 95, 105, 106, 181 and 184 as promising, ones.

Physico-chemical characteristic studies of fruits of five karonda lines *viz.*, PK1, PK4, Pant Manohar, Pant Sudarshan and Pant Suvarna indicated that fruit diameter, shape index, weight, volume, number of seeds per fruit, dry weight of flesh. fruit colour (Table 9.3), TSS and acidity differed significantly while fruit length, specific gravity, flesh, seed and ascorbic acid contents of fruits did not differ significantly among the selected lines.

Only few selections are made based on variation in fruit colour, size and quality. Work initiated in India during the 1980's on collection and evaluation of germplasm led to identification of a few location specific cultivars. Quality characters of ripe fruits of karonda lines PK-3, PK-4, Pant Manohar, Pant Sudarshan and Pant Suvarna were studied at Horticultural Research Station, GBPUAT, Pant Nagar at Patharchatta by Misra and Jaiswal (1998). Characters of selected lines of karonda developed by different institutions are enumerated below:

Pant Suvarna

This is a seedling selection at G.B. Pant University of Agriculture and Technology, Pantnagar. Plant grows upright and is bushy in nature. Fruits are green with light maroon blush. On maturity, they attain deep maroon colour. Fruit size is 2.55 cm x 1.66 cm and average fruit weight is 3.62 g. Pulp is 88.27 per cent, dry matter 12.39 per cent; Total soluble Solid 3.83 per cent, acidity 2.30 per cent and average yield/bush is 25 kg.

Pant Manohar

Pant Manohar is developed by a selection at G.B. Pant University of Agriculture and Technology, Pantnagar. Fruits are attractive in colour having deep pinkish blush on white background. Fruit is medium, size (2.13 cm x 1.69 cm) having an average fruit weight of 3.49 g. Pulp composition is 88.27 per cent, dry matter 12.77 per cent, total soluble solids 3.92 per cent and acidity 1.82 per cent. Average yield/ bush is 35 kg.

Table 9.3: Physical Characteristics of Fruits of Selected Lines of Karonda*

Selected Lines	*Fruit Length (cm)*	*Fruit Diameter (cm)*	*Fruit Weight (g)*	*Seeds Fruit^{-1}*	*Flesh (per cent)*	*Seed (per cent)*	*Fruit Colour*
PK1	2.16	1.70	3.53	4.48 (69.10)	87.30 (20.90)	12.70	White-pink
PK4	2.24	1.71	3.74	5.61	87.85 (69.60)	12.15 (20.40)	Green-purple
Pant Manohar	2.14	1.70	2.50	3.95	88.28 (69.99)	11.72 (20.01)	White-dark pink
Pant Sudarshan	2.17	1.96	3.47	4.68	88.48 (70.14)	11.52 (19.86)	White-pink
Pant Suvarna	2.25	1.67	3.63	3.89 (69.96)	88.25 (20.04)	11.75	Green-purple
S.E.m±	0.02	0.01	0.11	0.20	0.43	0.40	
C.D. 5 per cent	NS	0.03	0.30	0.60	NS	NS	

Figures in parentheses indicate transforme values.

Pant Sudarshan

It is a medium sized bush. Fruits are light pink measuring 2.16 x 1.69 cm. Average fruit weight is 3.46 g and seeds/fruit is 4.68. Pulp is 88.47 °A, dry matter 11.8 per cent, Total soluble solids 3.45 per cent, sourness 1.89 per cent and average yield is 32 kg $bush^{-1}$.

Maroon Coloured

Identified as a promising type developed at N.D. University of Agriculture and Technology, Kumarganj, Faizabad has an attractive maroon coloured fruit. Average fruit weight is 3.2 g, T.S.S. 7.1 per cent, acidity 4.1 per cent, vitamin C 17.9 mg $100g^{-1}$, anthocynanin (0D) 0.7, Iron 3.8 mg 100 g^{-1}. Average yield is 3.7 Kg $bush^{-1}$.

White-Pink Blush

This is also developed at NDUAT, Faizabad. Fruit weight is comparatively lesser than Maroon coloured 2.9 g. Average seeds $fruit^{-1}$ is 7.0, T.S.S.- 6.6 per cent, acidity 4.7 per cent, vitamin C 15.4 mg 100 g^{-1}, anthocyanin (OD) 0.2 and iron 4.15 mg 100 g^{-1}. Average yield $bush^{-1}$ is 2.8 kg.

Konkan Bold

Average yield 3.5 kg $tree^{-1}$ year $^{-1}$. It is early bearing, bold fruit size variety, developed by Dr. Balasaheb Sawant Konkan Krishi Vidyapeeth,Dapoli (Mahrashtra) during 2004.

Besides these promising karonda types, white and red fleshed karonda are located in tribal tract of Chattishgarh, Bastar and are under evaluation at College of Agriculture, Jagdalpur, Bastar and Chattisgarh.

Root Stocks

Shankar (1966) reported that C. *carandas could be* successfully grafted on C. *carandas, C. grandiflora* or *C. bispinosa* root stock and C. *grandiflora* on C. *grandiflora* rootstock. Success was obtained with C. *grandiflora* on C. *carandas* root stock but grafts of C. *grandiflora* on C. *bispinosa* rootstock did not survive.

Propagation

A. Seed

The most common method of propagation of karonda is by seeds. Fresh seeds should be used for raising seedlings since seeds usually lose their viability within 4-5 weeks. The germination of seed starts 40-45 days after sowing and is completed within 60 days. Bankar (1987) obtained early and enhanced germination (68 per cent) with increased vigour of seedlings when seeds were soaked in 100 ppm GA_3 for 24 hrs Seedlings are raised in polythene bags during rainy season. The seedlings are slow growing and hence they are ready for transplanting after about two years from sowing of seeds.

Seed has short viability, hence should be sown just after extraction from ripe fruits. Ideal time for sowing is rainy season. Seeds treated with GA_3 (25 ppm) improve seed germination (67 per cent) 58 days after treatment (Bankar, 1987). One year old seedlings are used for planting preferably after first shower of monsoon. For promoting early growth, gibberlic acid (200 ppm) have been reported effective by Misra and Jaiswal (1998).

The soaking of seed of Karonda in 20 ppm IBA for 12 hour shown the highest seed germination 80 per cent and seedling growth in respect of No. of leaves, seedling girth and height of seedling reported by (Raut *et al.*, 2011).

B. Vegetative Propagation

In respect of vegetative propagation in karonda, hard wood cuttings from trimmed hedge resulted (20 per cent) rooting during November. Cuttings from well established mother plants under nursery gave the best result. Karonda cutting rooted 10 per cent in late September, 20 per cent in early October, 30 per cent late October and 50 per cent in early November. In all cases, cuttings were pre-treated with Indole butyric acid at 500 ppm in 50 per cent alcohol. Shankar (1962) reported air layering for propagation of 3 species of *Larissa* and reported rooting in *C. carandas, C. grandiflora* and *C. hispanic* by treating with IAA and IBA at higher concentrations under Allahabad conditions. Soft wood grafting is also successful and a very useful technique for *in situ* propagation of karonda in arid zone (Chundawat, 1990).

In order to overcome the disadvantages of seed propagation, karonda can also be propagated by vegetative methods. It ensures the perpetuation of promising types. Interspecies grafting in genus *Carissa* has been reported by Shankar (1966).

1. Cuttings

Karonda is known to be difficult to root plant, therefore, rooting does not occur easily. Attempts were made by Mehata and Singh (1960), to propagate karonda by stem cuttings. Cuttings from young plants and hedge sources proved to be the best for rooting and shoot growth. Divekar (1984) reported more rooting with 1000 ppm IBA treatment as quick dip. No rooting was obtained without growth regulator treatment. The cuttings are usually taken from mature, about one-year-old shoot and should be 30-35 cm long with six leaves. Rainy season was observed to be the best period for propagation by cuttings. Treatment with 3000 ppm IBA and keeping the cuttings under mist markedly improved root formation. After 6-8 months, the plants attained transplantable height.

2. Air Layering

Jauhari and Nigam (1958) found high percentage of success with air layering when layers were treated with mixture of 7500 ppm each of IBA and NAA. Mishra and Singh (1990) reported that all the air layered shoots treated with IBA and NAA at 2500 to 10000 ppm in lanolin paste rooted whereas, untreated layers failed to root.

The best season for air layering is the beginning of monsoon i.e., June-July, it can be extended further upto September and the air layers should be treated with 4000 ppm IBA to get promising results (Divekar, 1984). Rooted layers were obtained after 3.5 month of layering.

Observation of air layered for days required for rooting, rooting percentage, number of roots, length of root, leaf tall, height of rooted layered, number of leaves, leaf area, root volume, root shoot ratio, dry weight of root, fresh weight of root and survival percentage of air layers were significantly superior when treated karonda layers by 5000 ppm IBA concentration. The final survival percentage was significantly superior 90 per cent in IBA 5000 ppm followed 82 per cent under IBA 6000 ppm.

3. Budding

Propagation of karonda by shield budding is economic and convenient. Natal plum and promising types of karonda could be budded on one-year-old rootstocks of C. *caronda* (Bankar *et al.*, 1994). The best time for budding is July to August in low rainfall areas.

4. Inarching

Propagation of karonda by Marching also proved successful. However, compared to other methods, inarching is not very popular (Bankar *et al.*, 1994).

5. Soft Wood Grafting

Soft wood grafting is also successful in karonda and is a very useful technique for *in situ* grafting in arid regions.

Karonda is commonly propagated by seeds and vegetative methods like stem cuttings, inarching and air layering. The fresh seeds are sown in nursery in the month of August or September. Under good management, seedlings become ready in a year for transplanting. Karonda can be propagated by hard wood cuttings successfully. Though cuttings are difficult to root type, they may also be planted under polythene tent which gives higher rooting because under such structure the humidity is high. Where mist is available, soft and semi-hardwood cuttings can be planted for successful rooting and rooted cuttings are transferred to pots containing good growing media consisting two-third of garden soil and one third of well-rotten compost. After 6 to 8 months, plants attain transplantable height. Soft wood grafting is also successful in karonda and is very useful technique for *in situ* propagation in arid regions. Air layering is quite successful in karonda and it is done in the beginning of monsoon. Rooted layers are obtained after 3 and ½ months of layering.

6. Micropropagation

Shoot tips from field grown; mature plants of *Carissa carandas cv.* Pant Sudarshan were cultured on Murashige and Skoog's (MS) basal medium supplemented with benzyl adenine (BA) and indole -butyric acid (IBA) during different seasons. The maximum sprouting rate was obtained with 1.5 cm long explant collected in spring season (February–March) followed by those collected in summer season (April–June). Shoot proliferation was the highest on MS basal media supplemented with 3.0 mgL^{-1} BA. Rooting of micro shoots was noted to be the best in ½ MSplus 0.8 mg L^{-1} IBA and 0.2 mg L^{-1} Naphthalene acetic acid (NAA). The rooted plantlets were successfully acclimatized in vermiculite:sand:soil (1:1:1) potting mixture reported by (Ratna and Misra, 2003).

Cultivation Practices

1. Planting

The field is cleaned, thoroughly ploughed and levelled. The pits of 30 to 45 cm^3 are dug and filled with organic manure and soil in the ratio of 1: 1. The distance of planting in hedge is recommended at 1 to 1.5 m and a regular planting at 1.5 to 2.0 m both ways. The best time for planting is the beginning of monsoon.

2. Planting System

Planting should be done in square system at 2m x 2m distance or hedge row plantation along boundary of orchard at a distance of 1m x 1m at prefilled pit (50 x 50 x 50 cm) with organic manure and soil (1:2).With onset of monsoon, planting should be done as per spacing. Total number of plants per hectare is 2500 when planted at a spacing of 2 m x 2 m. Whereas, 400 plants ha^{-1} are required when planted along boundary at a spacing of 1 m x 1 m. Planting should be done during monsoon, particularly in dry regions. This reduces need of frequent watering until plants are firmly established in soil. Karonda needs minimal after care however, hoeing of soil around plants, once a year helps better plant growth.

Karonda orchard can be laid out in square or rectangular system. Hedge planting can also be practiced by planting at 60 cm distance in double rows along the boundaries of the orchard. The pits of 30 to 45 cm are dug and filled with organic manure and soil in the ratio of 1: 1. The distance of planting or hedge purpose is recommended at 1 to 1.5 m and regular planting at 1.5 to 2 cm both ways. The best time for planting is the beginning of monsoon.

3. Training and Pruning

At the time of planting each plant is supported with the help of wooden stake so that it remains errect. Unwanted laterals are removed from time to time. Bearing tree, usually do not require pruning. However, to give desired shape and making the stem clear, additional twigs are pruned. Twigs which are affected with disease or broken are removed. Old hedge may be headed back to induce new growth.

4. Irrigation

Karonda requires irrigation during March to June. This is also the period of flowering in karonda. Therefore, the plants should be irrigated during summer at 10 to 15 days interval. In young plants, the light and frequent irrigation during summer are more useful. It does not require any irrigation during winter, except when there is danger of frost.

Karonda is a slow growing plant. Once established, it will not need much water. Moreover, it behaves as a xerophytic plant and therefore, water requirement is very low. However, efforts should he made to collect run-off during, rainy season in root zone through land shaping or water harvesting models.

Owing to its xerophytic nature, karonda plant needs very little irrigation after establishment. If there is no rain during period of fruit development, one irrigation is given for increasing fruit size. Irrigation after new planting and manuring in grown up plant is beneficial. In arid regions, water harvesting structure should be created to collect runoff water during rainy and post monsoon season near root zone for proper growth and development of plants.

5. Fertilizer Application

In the early years of the planting, plants require plenty of nitrogen for its growth. The fertilizer should be applied at bimonthly intervals after dividing the fertilizer in six parts. The fertilizer requirement of karonda are as below:

Table 9.4: Fertilizer Requirement of Karonda Plant

Sl.No.	*Age of the Plant (Years)*	*Urea (g plant^{-1})*	*Single Super-Phosphate (g plant^{-1})*	*Muriate of Potash (g plant^{-1})*
1.	1	100	150	75
2.	2	200	300	150
3.	3 and bove	300	450	225

Phosphorus and potassium fertilizers should be applied after soil testing. In most cases, it is advisable to supplement the fertilizer applications with micronutrient sprayings. Make three applications of limiting micronutrients.

About nutrition of karonda, research based data on the fertilizer doses are not available. It has been observed that this plant responds better to organic matter and an application of 10-15 kg FYM per bearing plant per year is suggested in the beginning of monsoon.

Being an underutilized fruit crop, very meagre information is available on manure and fertilizer requirements. In practice, dose of manure and fertilizer for optimum growth and bearing of karonda plants depends upon nutrient status of orchard soil. As a matter of common experience, it is observed that crop responds well to organic manuring. Without adding optimum quantity of manures and fertilizers, orchard soils slowly get exhausted and ultimately karonda plants show poor vegetative growth, produce fruits of poor quality and subsequently the die back symptoms appear. Well rotten Farmyard manure @ 10 kg plant^{-1} with onset of monsoon will be beneficial in dry areas. Application of 50:100:150 g plant^{-1} year^{-1} of N: P_2O_5 : K_2O respectively should be done for better results (Pandey and Reddy, 1989). Srivastava (1984) suggested application of 2-3 kg well rotten FYM and 100 g of fertilizer mixture of N P K to each of bearing plant for better yield.

6. Canopy Management

Karonda can be trained on single or double stem in regular plantations. For proper training, lateral shoots should be removed upto 15-20 cm height from ground level. This forms main stem of plant. Suckers arising out from ground, diseased, insect infested and dried twigs should be removed at regular intervals. Plants grown as protective hedges around other perennial orchards do not require any pruning except old hedges which may be headed back to induce new growth and form thick, impenetrable barrier for common animals. Bearing plants usually do not required pruning. To give desired shape and keeping stern clean in young unwanted twigs should be pruned as and when required during preparation frame.

7. Weed Management

Although weeds are not a major problem in karonda, but in general, field should be free from weed at initial stage of growth. Elmore *et al.* (1979) that oxyflurofen 2 per cent granules at 1.12 or 2.24 kg ha^{-1} or oxadiazon 2 per cent granules at 2.24 kg ha^{-1} gave good control of *Oxalis corniculata, Euphorbia maculata* and *Senecio vulgaris* in karonda orchard. Oxyflurofen treatment caused necrosis of young growth of *C. grancliflora* hence, one has to be careful in using herbicides in young karond orchards for weed management. Spray of 1.25 kg ha^{-1} oxadiazon and application of 0.6 kg ha^{-1} of 2 per cent Oxyflurofen granules are effective in weed management. (Brosh and Zilberstin, 1985). Considering ecological impact of pesticides on soil properties, it is felt that till the agricultural labour is relatively cheap and abundantly available, manual weeding should be given preference over chemical weed control practices. Exposure of soil to direct and strong sunlight by deep summer ploughing could be another alternative and most effective method of weed control, more so under fragile agro-ecosystem and with resource poor farmers.

8. Intercropping

In the first year of planting weeds may pose a problem which can be very effectively controlled by hoeing. In the regular plantation of karonda many seasonal vegetables can be grown in the first two years. They should be grown only during rainy season.

Flowering, Pollination and Fruit Set

Hooker (1882) described that the genus *Carissa* produces perfect flowers terminally or axglary. The calyx consists of five sepals and the corolla forms cylindrical tube. The stamens are present at the tip of corolla tube. Rendle (1925) reported that flowers are solitary and present in racemose in florescence passing into cyme. This species of *Carissa* is divided into two groups, one with short styled pistil and functional anthers and another with long styled pistil and small anthers devoid of pollen Heterostyly results in low production because presence of a high proportion of short styled flowers which are unfruitful.

Hains (1910) reported that flowering in karonda starts from April -May and continues upto December under Chhota Nagpur (Bihar) conditions. *Carissa grandillora* produces flowers and fruits throughout the year but the quantum of flowers was less during winter months. Bhagwat (1984) reported that flowering takes place from December to March, stigma is receptive upto one day after anthesis and the fruits ripen in a about 70 days after fruit set in Konkan region of Maharashtra. According to Misra *et al.* (1968) flower bud differentiation occurred on 7th February, when temperature was 29°C and it was completed by 14th March.

During this period, the average temperature ranged from 30.5 °C to 33.8 °C. A broadening and flattening of apical meristem with two protuberances one on either side of it was established as criterion of blossom bud differentiation. Karate *et al.* (1990) reported a large variation in average number of flower buds per umbel (14.6 to 22.6), length of flower bud (1.42 to 2.15 cm) and length of pedicel (0.7 to 1.32 cm), petal number (4 to 6) and flower groups i.e. umbel segments (1 to 6 in number). The fruit set under natural or open pollination was higher (32.02 per cent) than in self-pollination (24.24 per cent).

Fresh pollen grains were fine powdery white mass when observed with naked eyes. Size of pollen grains ranged from 29.1 to 37.5 microns and 15 per cent sucrose was found better medium for pollen germination (Singh *et al.*, 1989).

Studies on flowering and fruiting in *C. edulis* at Jodhpur revealed that it has two fruiting seasons. One in January-February and another in August-September (Singh *et al.*, 1982). Tiny flower bud takes about 19 days for flowering. The anthesis is between 12.30 to 4.30 p.m. Only 36 per cent of the flowers developed fruits to maturity (Banker and Verma. 1987).

Karonda starts bearing flowers and fruits from third year of its planting. There is no problem of pollination in karonda. The plant starts bearing flowers in the month of February and ripe fruits are available in the month of August. However, unripe fruits are available from May onwards. Under arid conditions, its flowering will coincide with monsoon and fruits will be available in post monsoon period.

Insect Pests and Diseases

As in other crops, here too the pests and diseases reduce the yield.

Plant Protection

Insect Pest Management

There is no serious pest infesting karonda plants. During initial stage of crop growth, it is infected by larvae of *Diggama hearseyana*, which results in stunted growth due to defoliation of leaves. This can be controlled by spray of 0.1 per cent nuvocron. Agrawal (1984) reported karonda plants to be infected by *Simcronyx roridus*, a gall forming weevil almost throughout year with its peak from March-May. Removal of infested twigs is the most practical approach of managing gall midges in karonda.

A. Pests

In the early stages, leaf eating caterpillar feed voraciously on the leaves and cause mortality of the plants. These caterpillars can be controlled by spraying Nuvacron @ 1 mL litre^{-1} of water.

Disease management

Several diseases infect karonda plant. Major diseases are Anthracnose (*Collectotrichum inamdrii*) and Bacterial canker (*Xanthomonas carissa*).

B. Diseases

Among the diseases, anthracnose and bacterial canker are of common occurrence.

Anthracnose

This is a fungal disease caused by *Colletotrichum inamdarii.* (Lal and Singh, 1953). This disease is widespread in Uttar Pradesh and can be seen in the form of black grey spots on the leaves. Spray of copper sulphate @ 0.1 per cent can control this disease.

This disease is of wide occurrence, more especially in humid and high rainfall areas. The leaves, shoots, inflorescences and the fruits are affected by it. The affected young shotts finally show dieback symptoms. The diseased young fruits drop down and if the fungus attacks the fruits at maturity, the disease is enhanced by storage. Spraying with copper sulphate (0.1 per cent) or blitox or phytolan (0.3 per cent) or Bavistin (0.1 per cent) or benlate (0.1 per cent) thrice a year is recommended (Lal and Singh, 1953).

Bacterial Canker

This is a bacterial disease and caused by *Xanthomonas carissa.* Symptoms first appear as spots on the lower surface of the leaf. These are small, round and water soaked which turn dark brown. A yellowish green halo may surround the spots which causes necrosis. The disease can be checked by removing diseased leaves followed by a spray of Phytomycin (200 ppm).

This disease is wide spread in Rajasthan. Twigs show similar spots. Flowers and fruits do not show any symptoms (Kumar and Patel, 1968). At Jobner, Rathore (1970) observed maximum number of leaf lesions in third week of October (24°C and 56 per cent RH). The disease can be checked by removing all the diseased leaves from the shrub pruned to the height of 45 cm from the ground level. Spraying with phytomycin (200 ppm) can also control the disease (Prasada *et al.*, 1971). Leaf blotch of *karonda* caused by *Pseudocercospora* was observed by Singh and Mukherjee (1979) under Faizabad conditions.

Post-harvest Diseases of Karonda Fruits

Fruit rot of karonda due to *Myrothecium roridum* (Kumar and Tandon,1979) and brown rot of fruits due to *Alternaria alternata* (Tandon and Kumar, 1974) were observed as common and serious disease in local orchards in Allahabad conditions of Uttar Pradesh. These diseases rendered fruits unfit for human consumption. Both immature and mature fruits are susceptible. Fruits of karonda are also attacked by fungus *Bartalinia robillardoides* Tassi (Sharma and Agrawal, 1976). Beacause of infection of fruits with fungi like *Pestalotiopsis versicolar, Myrothecium roridum* and *Alternaria alternata* quantity of tartaric and malic acid decrease while fumaric acid disappears completely in infested tissues (Kumar and Tandon,1979).

Harvesting and Yield

Harvesting

Plants raised with seeds start bearing in third year of its planting. The plants flower during March. The fruits ripen from July to September in North India. In arid conditions, flowering starts late and fruits ripen in post monsoon period. Karonda requires 2-3 pickings. On an average a plant provides 3-5 kg fruits.

Karonda fruits mature 100-110 days after fruit set. At this stage fruits develop their natural colour. Fruits ripen after this stage, taking, about 120 days (after fruits set) when they become soft and attain dark purple/maroon/red colour.

Pawar (1988) concluded that weight, volume, specific gravity, length, diameter and colour of the fruit can be considered as the physical indices whereas, moisture, TSS, sugars, acidity, pH and tannins as chemicals indices of maturity of karonda fruits. There is no standard practice for grading and packing of fruits. Fruits after harvesting are kept in shade. Undesirable or blemished fruits are sorted out. Good fruits packed in baskets are marketed. Storage life of fruits depends upon the stage of harvest. Traditionally, the fresh fruits are collected by tribal people from forest areas and marketed in fresh condition.

Flowering starts in March and fruits ripen from July onwards upto September with coincidence of rainy season. Unripe fruits are available from May onward. Change of colour with specific gravity above 1.0 is a good indicator of maturity of fruit. Two to three pickings are generally done for a complete harvest of 4-5 kg fruits plant^{-1}. In sandy soils, 5 kg fruit tree^{-1} year^{-1} can be harvested under normal conditions. Under good management conditions, 10-12 kg fruit yield from a full grown tree year^{-1} is reported by Pareek (1990). Singh *et al.* (1982) reported yield of 500-600 number of fruits^{-1} hush^{-1} year from *C. grandiflora* and 1.2-1.6 kg from *C. mulls* under Jodhpur conditions from 7-8 year old trees. Fruit yield varied from 13 kg plant^{-1} under hedge row planting system along field boundary in Agra conditions (Kumar and Yadav, 2006).

Karonda fruits should be harvested when they attain proper maturity. Both ripe as well as unripe fruits are harvested. Complete harvesting for all the fruits may not be done at the same time. Harvesting is normally done two to three times. Colour change is a good indicator of maturity of fruits. About 4 to 5 kg fruits per bush is obtained.

Storage and Post-harvest Technology

Karonda has also remained neglected as far as its post harvest handling and technology is concerned. Generally, the tree-ripe fruits are harvested. Because of its soft flesh with high moisture content, its shelf life is short.

Fruits harvested at maturity, can be stored for a week at room temperature, whereas, fruits harvested at ripe stage are highly perishable and can be stored only for 2-3 days. Fruits can be preserved/stored for 6 months in SO_2 solution (2000 ppm). The fruits harvested with stalk intact recorded less shrivelling and microbial spoilage (Pawar, 1988). For reduction of spoilage, the cold storage (13°C and 94 per cent RH) could be considered the best while for reduction of shrivelling, cool chamber could be more effective.

Fruits harvested at maturity can be stored for 5-7 days at room temperature before they begin to shrivel. If harvested at ripe stage, it has a very short life and can be stored only up to two days at normal temperature. Fruits can also be stored for long time in a solution of Potassium Metabisulphite (2000 ppm).

Processing for Value Added Products

The products such as squash, syrup, jam, jelly, pickle, chutney, dried fruits and candy can be successfully prepared from the mature green and ripe fruits. In southern Florida, ripe karonda fruits are utilized for processing into syrup and wine (Sturrock, 1950). Nalawadi (1967) has reported that preparation of attractive pink coloured jelly from ripe fruits.

Chutney

Ingredients required for making chutney from 1 kg fruit is 50 g chilli powder, chopped ginger, salt 10 g, garlic chopped 10 g, hot spices 20 g and acetic acid 20 mL. Fruits after thoroughly washing are cut into slices and boiled with a little water by adding sugar and salt. While cooking is continued, chopped ginger in a muslin bag are put in it and pressed with a spoon to obtain the extract. After cooking to a desired consistency spice bag is again slightly heated and pressed and thereafter, it is filled in clean bottles, capped and stored in cool and dry place.

Pickle

For making pickle, after washing fruits, salt (200 g) is added and kept for 3 days. On fourth day all spices (chilli powder 25 g, mustard powder 50 g, turmeric 10 g, fenugreek 10 g) is added and mixed thoroughly. After mixing, 300 mL mustard oil is added and filled in jar followed by capping and storage in a dry place.

Jelly

Karonda jelly is made by boiling mature fruits. Fruits after washing in water are cut in the form of thin rings and equal amount of water is added. After adding water, it is boiled for 20-30 minutes and 2 g citric acid is added. After boiling is completed, extract is strained and tested for pectin content. Thereafter, 1 kg sugar is added and cooked and during this cooking process, scum is removed. Finally, the ready product is filled into clean and sterilized bottle followed by sealing, capping and storage in cool and dry place.

Candy

Mature and healthy fruits of karonda are washed and seed is removed by giving a slit with help of a sharp knife. The ready fruits without seeds are steeped in 2 per cent salt solution containing Potassium Metabisulphite (KMIS) (600 ppm) for 24 hours to bleach fruits followed by washing, pricking and blanching it water containing 0.05 per cent erythrosine and 0.25 per cent citric acid to fix colour. Thereafter, fruits are steeped in sugar syrup of 50 per cent for 24 hours

Karonda fruit (mature)

Washing

↓

Cutting in to two pieces

↓

Removal of seeds

↓

Mixing with salt

Filling in jar boottles

↓

Keeping in sun for 4 days

↓

Frying spices in oil

↓

Mixing fried spices with cut karonda fruit pieces

Sealing of bottles

↓

Storage in cool and dry place

Figure 9.1: Flow Chart for Preparation of Karonda Pickle (Srivastava and Kumar, 1994).

and subsequently to higher concentration of sugar syrup in phased manner till it is steeped in 75 per cent TSS sugar syrup. Adhered syrup is ultimately removed by dipping in boiling hot water for few seconds. The ready product is packed in suitable bottles, sealed and stored in cool and dry places.

References

A. Zaki, S. El-Tohamy, S. El-Fattah. Study of Lipid content and volatile oil of the different organs of *Carissa carandus* Lin. and *Carissa grandiflora* Dc. growing in Egypt. *Egyptian Journal of Pharmaceutical Sciences* 22(14):127-41 (1983).

B. Singh, R.P. Rastogi. The structure of carindone. *Phytochemistry*, 11(5):17 97-801 (1972).

Chunekar, K.C. Bhavaprakasa Nighantu, Chaukhambha Bharati Academy, Varanasi, (1982).

D.V. Joshi, S. F. Boyce. Chemical Investigation of Root of *Carissa congesta*, Santapau. *Journal of Organic Chemistry*.22: 95-7 (1957).

Hegde K., Thakker S.P., Joshi A.B., Shastry C.S., Chandrashekhar K.S.. Anticonvulsant Activity of *Carissa carandas* Linn. Root Extract in Experimental Mice, *Tropical Journal of Pharmaceutical Research.* 8 (2): 117-125 (2009).

Indradev Tripathi. *Raja Nighantu,* 2nd edition, K. Gadas Academy Chaukhamba Press Varanasi, 1998.

J. Pino, R. Marbot, C. Vazques. Volatile flavour constituents of Karnda (*Carissa carandas* L.) fruit. *Journal of Essential Oil Research.*16(5): 432-4 (2004)

J. Reisch, R. Hussain, B. Krebs, M. Dartmann. The structure of carissone. Monatshefte fuer Chemie 121(11): 941-4 (1990)

K.L. Noatay (2004), "Karonda, the pickle berry". *India Tribune,* retrieved June 4, 2007.

M.M. Vohra, De N.N. Comparative cardiotonic activity of *Carissa carandas* (L.) and *Carissa spinarum {A}. Indian journal of medical research.* 51(5): 937-940 (1963).

Morton, J. 1987. Karanda. p. 422–424. In: Fruits of warm climates. Julia F. Morton, Miami, FL. Shashi K. Sharma and S.D. Badiyala (2002), "Karonda, fruit plant for marginal lands", *India Tribune,* retrieved June 1, 2007.

Nagaur, Rajasthan Department of Zoology, Ganjdundwara P.G.College, Ganjdundwara.

Parveen, B. Upadhyay, S. Roy, Kumar A. Traditional uses of medicinal plants among the rural communities of Churu district in the Thar Desert, India. *Journal of Ethnopharmacology* 113: 387-399 (2007).

Shashi K. Sharma and S.D. Badiyala (2002). "Karonda, fruit plant for marginal lands".

Raaz Maheshwari, Anju Sharma And Deepmala Verma Department of Chemistry, Sbrmgc.

R. Pal, D.K. Kulshreshtha, R.P. Rastogi. A new lignan from *Carissa carandas. Phytochemistry.* 14(10): 2302-3 (1975).

S. Siddiqui, U. Ghani, S. Ali, S. Usmani, S. Begum. Triterpenoidal Constituents of the leaves of *Carissa carandas. Natural Products Research.* 17(3):153 (2003).

10
Jamun

A.B. Bhosale, B.D. Tamboli and S.M. Todmal

Department of Horticulture,
MPKV, Rahuri, Maharashtra

Jamun [*Syzygium cumini* (L.)], highly adapted to diverse environmental conditions, is widely distributed in India up to an altitude of 1800 m. It is cultivated commercially for fruit and timber and also for medicinal uses. Jamun is a tall handsome evergreen tree grown in tropical and subtropical regions. It is one of the hardiest fruit tree and best suited for wastelands. It is drought tolerant, at the same time can tolerate water stagnation as well as marshy wetlands. The wood is used as a timber in building and railway sleeper.

The nutrients contents out of 100 gm of fresh fruits are 19.7 g carbohydrate, 0.7 g protein, 0.1 g fat, 1.0 g iron, 0.02 g calcium, 0.01 g phosphorus and 0.9 g fibre. Besides taken as a dessert fruit, it is also used for making value added products.

Jamun botanically called as *Syzygium cumini* L. (Syn: *Eugenia jambolana, Syzygium jambolanum*) belong to the family *Myrtaceae.* The tree is originated from Indonesia and India; now grows abundantly in Southern Asia. It is an evergreen tropical tree having 8 to 15 m tall, with oblong opposite leaves that are smooth, glossy and having a turpentine smell. The bark is scaly gray and the trunk is forks or multiple stem, has fragrant white flowers in branched clusters at stem tips and purplish-black oval edible berries. This berry has only one seed. The juicy fruit pulp contains resin, gallic acid and tannin. The taste is usually acid to fairly sweet and somewhat astringent, jamun fruit mainly used for its medicinal properties.

Climate

Jamun is successfully grown under tropical and subtropical climate. It also occurs in the lower range of the Himalayas upto an elevation of 1300 metres and in the Kumaon hills upto 1600 metres above sea level. It is widely grown in the larger parts of India from the Indo - Gangetic plains in the North to Tamil Nadu in the South. It is one of the most hardy fruit crop and can be easily grown in neglected and marshy areas where other fruit plants can not be grown successfully. Jamun is somewhat more susceptible to cold and drought than the seedling mango trees. It requires dry weather at the time of flowering and fruiting. For ripening of fruit and proper development of its size, colour and taste, early rains are considered very beneficial. The fruits show remarkable improvement in these respects after the very first shower of rain.

Soils

Jamun tree is exact in its soil requirements. The tree requires deep, loam and well drained soils for its optimum growth and good fruiting. Its cultivation should be avoided in very heavy or light soils.

Choice of Varieties

Raw Jamun

It produces big sized fruit with average lenght of 2.5 - 3.5 cm and of diameter 1.2 - 2.0 cm. Fruits are oblong in shape, deep purple or bluish black in colour at fully ripe stage. The pulp colour of ripe fruit is purple pink and the fruit is juicy and sweet. The stone is small in size. It ripens in the month of June-July. The variety is very common among the people.

Small Sized Jamun

It is a late maturing variety. The average lenght of fruit is 1.5 - 2.0 cm and diameter is 1 - 1.5 cm. The fruit is slightly round in shape, deep purple or blackish in colour at full ripe stage. The colour of the pulp is purple, less in juice, weight and sweetness of pulp in comparison of that of 'Raw Jamun'. The stone is very large. Fruits ripen in the month of August.

Propagation and Rootstock

Though lot of jamun plantation is seen on road side or scattered plantation at farmer's field, yet no single well established orchard is reported in North India. Also no systematic work has been done on its propagation. The most common method of jamun propagation is by seed. Seedling plant bears fruit of variable size and quality. For improved and selected true-to-true, vegetative methods of propagation like inarching and air layering have been advocated.

Jamun is propagated both by seeds and vegetative techniques, the most common being by seeds. The seeds have no dormancy, hence fresh seeds can be sown (with in 10-15 days) 4-5 cm deep at a distance of 25 cm × 15 cm. The seeds germinate 10 -15 days after sowing. The seedlings become ready for transplanting in spring or next monsoon. Its seeds show polyembrony upto 30- 40 per cent, hence nucellar/ apomictic seedlings can be used to produce true- to- type plants.

Sexual Propagation

Seeds are sown fresh in flat nursery beds during July when this fruit ripens. The seeds should be sown at a distance of 15 cm in rows which are 25 - 30 cm apart. The seedlings can also be raised in polythene bags of 22.5 - 30 cm size. The bags should be filled with a mixture of soil and farmyard manure in equal proportion. To drain out excess water the polythene bags should be pricked from all sides before filling the mixture. It has been seen that more than one seedling comes out from a single seed. These seedlings are separated in different bags when they are about two weeks old. Plants grown from seed become transplantable during next spring season. But it is advisable to keep them in nursery upto next rainy season which is the best time of its plantation.

Vegetative Propagation

Seedlings plants bear fruits of variable size and quality. Therefore, vegetative method is desirable for propagation of improved or selected types. Budding is most successful for commercial rising of plants. It is done on one-year-old rootstock having about 10 mm thickness. Patch and Forkert methods give more than 80 per cent success if performed in March in low rainfall area; July- August is ideal time for budding.

T-budding and Patch Budding

According to the recent research conducted at Punjab Agricultural University has revealed that jamun can be the best propagated through T-budding as well as through patch buddding. The seedling of jamun is used as a rootstock. The per cent success is higher in T-budding (70 per cent) than patch budding (60 per cent). The best time for budding in both the cases is either during February-March or during August-September. However, the success during August-September is higher.

Inarching

The rootstock used for propagating jamun is the jamun seedlings. For raising the rootstock, seeds are collected from healthy, vigorously growing and high yielding jamun trees. Seedlings are raised either in bed or in pots singly. In the month of June-July one or two year old rootstocks are inarched with the matching thickness of scion. Rootstocks are watered if necessary till the grafts are separated from the parent tree. The union will complete in a period of about six weeks.

Veneer Grafting

Veneer grafting gives 31 per cent success when one year old seedlings are used as rootstocks. The shoots are taken from spring flush and the method is employed in the month of July.

Planting Operation

Pits of 1m X 1m X 1m size are dug 10 m apart for seedlings trees and 7-8 m apart for budded plants in a properly cleaned field. Pit digging should be completed before the onset of monsoon or spring season. They should be filled with mixture of top soil and well rotten farm yard manure or compost in a 3:1 ratio. Monsoon season July to September is suitable season for planting. About 100 -150 plants are required for planting a hectare land.

Jamun can be transplanted during spring (February-March) or during monsoon (August-September). However, the later season of planting is considered better because the plants easily get established during the rainy season. The plants are transplanted with earth ball and are given irrigation till they get established. The size of the pit should be 1 m X 1 m X 1 m and these should be filled with a mixture of surface soil, silt and well rotten farmyard manure. The jamun is planted at the distance of 10 - 12 metres in square system, thus, accommodates 105 - 75 plants per hectare.

Flowering and Fruiting

The flowering in jamun starts in the first week of March and continues up to the middle of April. The trees are in full bloom in the second week of April. The inflorescence in jamun is generally borne in the axils of leaves on branchlet. The flowers are hermaphrodite, light yellow in colour. The maximum anthesis and dehiscence were recorded between 10 A.M. and 12 Noon. The pollen fertility was higher in the beginning of the season. The maximum receptivity of stigma was observed one day after anthesis. Jamun is a cross-pollinated fruit. The pollination is done by honeybees, house flies and wind. The maximum fruit set i.e. 32.6 - 36.0 per cent was obtained when pollination was done one day after anthesis. Thereafter, a sharp decline was observed in fruit set.

Three distinct phases of fruit growth in jamun are recorded. During the first phase (15 - 52 days after fruit set), the rate of growth was slow. In the second phase (52 - 58 days after fruit set), the rate of development was quite rapid and the third and last phase (58 - 60 days after fruit set) comprised comparatively slow growth with little addition of the fruit weight. The lenght and diameter of fruit showed a continues increase with advancement of maturity. The colour of jamun fruit changed from dark green at fruit set to light reddish colour at partial ripening and dark or bright purple at full ripe stage. The fruit took 63 days for complete ripening from fruit set. The ripe jamun had 76 per cent edible portion and 3: 1: 1 pulp to seed ratio. TSS and sugars followed an increasing trend, while tannin content followed a decreasing trend during growth and development.

Orchard Cultural Practices

Irrigation

During the initial year, the jamun plants required 8 - 10 irrigations in a year while bearing trees (full grown) required 4 - 6 irrigations during the summer months of May and June when fruits ripen. In the remaining period of the year, irrigation may be given when there are no rains or dry spell persists. During winter month, irrigation proves useful as it protects the plants from frost injury.

Intercropping

To supplement the income from pre-bearing period of jamun, intercropping should be practised judiciously. Intercropping also improved fertility of the soil. Fruit crops like peach, plum, guava, kinnow, kagzi lime, phalsa and papaya can be grown as filler trees. Such filler trees can be uprooted when the jamun trees starts bearing commercial crop. The leguminous crop like gram, peas, moong and mash can be grown successfully. In addition, intercrops of vegetables near established market may be taken with cauliflower, cabbage, knol-khol, radish, brinjal, turnip, carrot etc.

Manuring and Fertilization

During the pre-bearing period of jamun, a dose of 20 kg well rotten farmyard manure should be applied. To the bearing tree, 80 kg FYM per tree should be supplied annually for proper growth and fruiting. Sometimes in highly fertile soils, the plants produce profuse vegetative growth and fruiting is delayed. Under such conditions, the manures should not be given and irrigation should also be given sparingly and with held in September-October and again in February-March. This procedure will prove beneficial in fruit bud formation, flowering and fruit setting. Ringing and root pruning are also helpful.

Irrigation

During early stages of establishment, the plants must be watered periodically for better growth. The full grown and bearing trees, irrigation should be given once in 12-15 days. Care and periodically watering must be given during summer months. It will help for bud formation, fruit set as well as prevents fruit drop.

Aftercare

During early stages of crop growth an inter crop like cowpea, green gram can be grown upto 5-6 years, intercropping with seasonal vegetables also advisable. Jamun is a cross pollinated fruit crop hence keeping honey bee boxes 4-5 per acre is desirable for maximum fruit set and high yield.

Training and Pruning

Young plants need training for the development of frame work. Keep the main stem or trunk clean up to a height of 90 cm from the ground level by removing the basal branches and sprouts. Jamun do not require any pruning except removing diseased, dry and criss-cross twigs.

Jamun plants should be trained according to the modified leader system. Regular pruning is not required in jamun plants. However, in later years, the dry twigs and crossed branches are removed. While training the plant, the framework of branches is allowed to develop above 60 - 100 cm from the ground level.

Flower and Fruit Drop

In jamun, the flower and fruit drop start just after opening of flowers and continue up to maturity. About 65 per cent flower and fruit drop in the first five weeks and since then a maxium of 19 - 21 per cent flowers and fruits drop off up to maturity. Only 12 - 15 per cent flowers reach maturity. The flower and fruit drop are found at 3 stages. The first drop takes place during bloom or shortly there after. This proves to be the heaviest drop as about 52 per cent of the flowers drop off after 4 weeks from flowering. The second drop starts about 35 - 40 days of full bloom and apparently there is no difference between the developing and aborting fruits. The third drop takes place after 42 - 50 days of full bloom and continues till 15th July.

Control

The extent of flower and fruit drop in jamun may be reduced by two sprays of 60 ppm GA_3, one at full bloom and the other 15 days after initial setting of fruit.

Harvesting and Fruit Handling

The grafted jamun starts bearing after 6 - 7 years while the seedling one after 8 - 10 years of planting. The fruit ripen in the month of June-July. The main characteristic of ripen fruit at full size is deep purple or black colour. The jamun fruit is non climacteric in nature. The fruit should be picked immediately when it is ripe, because it can not be retained on the tree in ripe stage. The ripe fruits are picked singly by hand and in all cases care should be taken to avoid all possible damage to fruits. For harvesting, the picker climbs the tree with bags of cotton slung on the shoulder. The fruits of jamun is generally harvested daily and sent to market on the same day. The fruit is highly perishable and can be kept in good condition for about 2 - 3 days under ordinary conditions.

The average yield of fruit from a full grown seedling jamun tree is about 80 - 100 kg and from a grafted one 60 - 70 kg per year. The storage life of jamun fruit is 6 days at room temperature and 3 weeks at low temperature (9°C and RH 85 - 90 per cent) when pre-cooled fruits are kept in perforated polythene bags.

Insects Pests

1. White-fly (*Dialeurodes eugeniae*)

It damages the tree in all parts of India. Sometimes the fruits of jamun get wormy due to attack of fruit fly. Pests can be controlled by maintaining sanitary situation in the orchard. Pick up the affected fruits and burry them deep in soil. The area under the tree should be dug so that the maggots in the affected fruits and the pupae hibernating in the soil may be destroyed.

2. Leaf Eating Caterpillar (*Carea subtilis*)

It is reported to damage the plant in South India at Coimbatore. The insect infest the leaves and may defoliate the trees. The pest can be controlled by spraying 625 mL Roger® in 500 litres of water.

3. Squirrels, Parrots, Crows and Birds

The jamun fruits are also damaged by these squirrels, parrots, crows and birds. For keeping them away, beating of drum or flinging small dry earthern balls through a sling is useful.

Diseases

Leaf Spot and Fruit Rot

The disease caused due to *Glomerella cingulata*. Affected leaf shows scattered spots, light brown or reddish brown in colour. The affected fruits rot and shrivel. The disease can be controlled with fungicide like Dithane Z - 78 at 0.02 per cent or Bordeaux mixture (2 : 2 : 250).

References

Brito ESD, Araujo MCD, Alves RE, Carkeet C, Clevidence BA, Novotny JA. Anthocyanins present in selected tropical fruits: acerola, jambolao, jussara, and guajiru. *J Agri Food Chem*, 55: 9389-9394, (2007).

Faria AF, Marques MC, Mercadante AZ. Vinod Rathore, N S Shekhawat*, R P Singh, J S Rathore and H Dagla. *Indian Journal of Biotechnology* Vol. 3, April 2004, pp. 241-245

Helmstadter, "*Syzygium cumini* (L.) Skeels (Myrtaceae) Against Diabetes: 125 Years of Research," *Pharmazie*, Vol. 63, No. 2, 2008, pp. 91-101.

J. Giri, T. Sathidevi and N. Dushyanth, "Effect of Jamun Seed Extract on Alloxan Induced Diabetes in Rats," *Journal of the Diabetic Association of India*, Vol. 25, 1985, pp. 115-119.

Mitraniketan Krishi Vigyan Kendra, Vellanad- 695 543, Thiruvananthapuram. e-mail: rpherbals@yahoo.com, Kerala Calling, 2007.

"*Syzygium cumini* (L.) Skeels". The Plant List. Royal Botanic Gardens, Kew and Missouri Botanical Garden. Retrieved 31 May 2013.

Veigas J, Narayan M, Laxman P, Neelwarne B. Chemical nature, stability and bioefficacies of anthocyanins from fruit peel of *Syzygium cumini* Skeels. *Food Chem*, 105: 619-627, (2007).

Zhang LL, Lin YM. Antioxidant tannins from *Syzygium cumini* fruit. *Afr J Biotecnol*, 8: 2301–2309, (2009)

Rajesh Kumar, V.V. Ramamurthy and Gaurav Sharma. Checklist of insects associated with Jamun (*Syzygium cumini* Skeels) from India. *Biological Forum: An International Journal*, 2(1): 1-5(2010).

Shrivastava, R.P. and Kumar, S. (2009). *Fruit and vegetable preservation. Principles and practices*, IBDC, New Delhi.

Index

A

Abscisic acid (ABA) 35
Adriatic 147
Agri-horticultural land use system 41
Agro-climatic zones 1
Agroforestry 41
Air layering 168
Air temperature 113
Ajowan 9
Alternaria alternata 174
Alternaria brassicicola 45
Alternaria chartarum 45
Anacardium occidentale 2
Annona sqnomosa 111
Anthracnose 174
Anthrocyanin 58
Antioxidants 140
Aonla 89
Apple 7
Arecanut 6
Arnica granatum 57
Aromatic 4
Aspergillus hasus 45
Aspergillus niger 152
Azospirillum brasilense 32

B

Bacterial blight 77
Bacterial canker 174
Bactrocera zonata 125
Bahar treatment 72
Banana 6
Bartalinia robillardoides 174
Ber 15
Ber plant 19
Biodiversity 22
Biological control 127
Biopesticides 127
Black ischia 147
Bottle gourd 8
Brinjal 8
Brown Turkey 146
Brunswick 147

Budding 24, 115, 135, 169
Budding method 25
Bulb vegetables 8

C

Caesalpinioideae 131
Candy 176
Canopy management 171
Carbon monoxide 5
Carea subtilis 186
Carissa carandas 160, 164, 169
Carissa spindrum 164
Carrot 5
Cashew 7
Cassava 6
Caucasus 57
Celeste 146
Chhuhara 39
Chilli 8
Chip bud 116
Chrysanthemum 10
Chutney 176
Cicer arietinum 132
Citrus 6
Cladosporium zizyphi 45
Cleft Graft 116
Climate 17, 60, 112, 132, 148, 163, 181
Climate change 3, 5
Climate condition 3
Climate Requirement 10
CO_2 8
Cocoa 6
Coconut 8
Composition 104
Conservation agriculture 12
Coriander 9
Crop regulation 72
Cucumber 8
Cucurbitaceous vegetables 8
Cultivars 36, 62
Cultural practices 126
Cumin 9
Cuscuta spp. 45
Custard apple 111, 126
Cutting 115
Cuttings 135
Cycocel (CCC) 35
Cylindrocladium scoparium 152
Cytokinins 35

D

Diagnostic Recommendation and Integrated System (DRIS) 31
Dialeurodes eugeniae 186
Diseases 99, 152, 173
Drought 6
Dryland 41

E

Ecology 146
Economics 104
Ellagic acid 58
Emblica officinalis 89
Environmental impact 113
Ethylene (CEPA) 35
Exchangeable sodium per cent (ESP) 163

F

Ferrisa virgata 125
Fertigation 70, 121
Fertilisers 28
Fertilization 69, 185
Fertilizer 12, 120, 137
Fertilizer application 171
Ficus carica 143
Fig 2, 143
Flooding 5

Flower crops 9
Flowering 99, 113, 184
Flowers 10, 132
Foliar feeding 32
Food uses 152
French bean 10
Fruit borer 124
Fruit crops 5
Fruit drop 185
Fruit fly 125
Fruit growth 114
Fruit handling 186
Fruit rot 51, 187

G

GDP 4
Genoa 147
Germination 21
Gibberellins 35
Ginger 10
Glomerella cingulata 187
Gola 40
Grafting 115, 135
Grapes 7

H

Habitat 145
Harvest 151
Harvesting 103, 175
Harvesting stage 137
Heliminthosporium atroolivaceum 45
Heterographis bengalella 124
Himalayan region 4
Horticultural crops 3
Horticultural varieties 36
Horticulture 5
Hydrofluride 4
Hypoxylon hypomilltum 45

I

Inarching 169, 183
Indian gosseberry 89
Insect pest management 173
Insect pests 45
In-situ moisture conservation 43
Integrated nutrient management (INM) 59
Integrated pest management (IPM) 126
Intensity 33
Intercropping 98, 136, 172, 185
Interculturing 99, 119
Internal fruit necrosis 99
Irrigated areas 28
Irrigation 71, 97, 122, 137, 170, 184, 185
Isariopsis indica 45

J

Jamun 181
Jelly 176
Jogia 41
Juice concentrate 139

K

K_2O 71
Kaithali 40
Karonda 159
Kathapal 40
Konkan bold 167

L

Laddu 39
Latex 154
Leaf 161
Leaf eating caterpillar 186
Leaf spot 187
Leaves 154
Light 133
Lsariopsis indica 45

M

Mangifera indica 2
Mango 6, 10
Manures 28, 137
Marketing 103
Marseilles 147
Mechanical practices 127
Medicinal uses 155
Microbial inoculations 32
Micropropagation 169
Moisture 28
Mulching 98, 123
Mundia 41
Muskmelon 8
Myrothecium roridum 174

N

Naphthalene acetic acid (NAA) 35
Nigrospora oryzae 45
Nitrogen oxide 4
Nursery management 22
Nutrients 29
Nutrition 96, 119, 137

O

Oidium spp. 45
Ozone 11

P

Pant suvarna 165
Patch budding 183
Pestalotiopsis versicolar 174
Phoma hessarensis 45
Phyllanthus emblica 89
Physiological disorders 99
Pickle 176
Pitcher system 44
Plant growth regulators 34, 35
Planting 26, 95, 136, 170
Planting procedure 117
Plant protection 74
Pollination 172
Pollutants 11
Polyphenols 58
Polysaccharide 140
Pomegranate 6, 57, 78
Poona 148
Post-harvest technology 175
Potato 5
Propagation 19, 22, 61, 91, 115, 133, 149, 167, 182
Pruning 32, 66, 97, 123, 136, 150, 170
Pruning rime 34
Pseudostem 6
Punica granatum 62
Punicalagin 58
Punica protopunica 62
Purple genca 147

R

Rainfall 10, 18
Rainfed conditions 32
Rain water harvesting 12
Raw jamun 182
Renewable energy 12
Residue management 12
Rhamnaceae 15
Ridge gourd 8
Rodent management 128
Root 161
Root-knot nematode 152
Rootstocks 23, 167

S

Saissetia haemispherica 152
Sanauri-5 39
Season 150
Seed 167

Seedlings 23
Seed oil 154
Seed pretreatment 21
Seed products 140
Seed propagation 20, 115, 134
Seeds 132
Seed treatments 134
Seed viability 20
Sexual reproduction 22
Shoot growth 113
Small sized jamun 182
Soft wood grafting 169
Soil 19, 60, 91, 112, 133
Soil drainage 114
Soils 149
Soil temperature 113
Solanaceous vegetables 8
Spacing 65
Species 164
Spices 8
Squash 5
Squirrels 186
Stemphyliomma valpara 45
Storage 139
Sulphur dioxide 4, 5, 11
Sweet potato 6
Syzygium cumini 181

T

Tailed mealy bug 125
Tamarind 131
Tamarind beverage 139
Tamarind pulp 139
Tamarindus indica 131
Tannins 58
Taxonomy 162
T-budding 116
Temperature 5, 18, 113, 132
Tikadi 40
Timber 4
Tomato 5
Topworking 117
Toxicity 154
Training 66
Triticum aestivum 132
Tropical tuber vegetable crops 8

U

Umran 39, 40
Unfruitfulness 103

V

Value added products 176
Varieties 91, 114, 146, 148, 165
Vegetative propagation 23, 95, 115, 168, 183
Veneer graft 116
Veneer grafting 183
Vitamins 15

W

Water 10, 132
Water harvesting 27
Water logging 5
Water management 43
Watermelon 5, 8
Weed control 123
Weeding 28
Weed management 127, 172
Whip graft 116
White-fly 186
White-pink blush 167
Wind 133
Wood 162

Z

Ziziphus jujuba 18
Ziziphus mauritiana 15, 18

Ber Fruits

Buddiing Method in Ber

1. Budwood

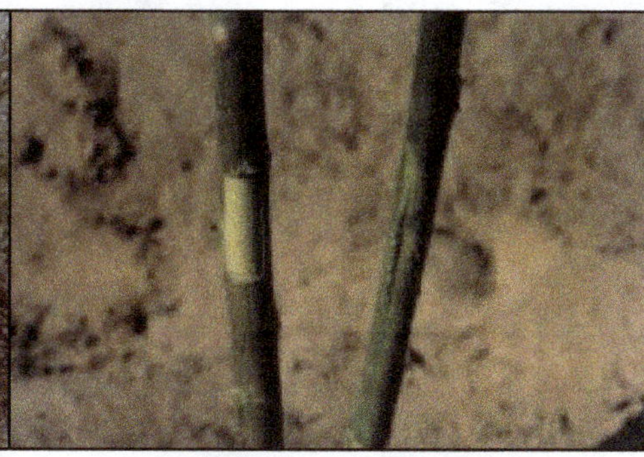

2. Rootstock prepared for patch or shield budding

3. Scion buds tied after insertion

4. Scion buds successfully sprouted

p. 53

Training in Ber

1. First year age ber plant

2. Second year age ber plant

3. Three year age ber plant

Pruning in Ber

p. 54

Moisture conservation in Ber

1. Mulching

2. Semicircular basin for *in situ* moisture conservation

p. 55

Pomegranate Varieties

1. Ganesh

2. Maskat

3. Phule Bhagawa

p. 87

Fertilizer application in Pomegranate

Mulching for moisture conservation

Aonla tree

Aonla orchard

Aonla fruit

p. 108

Aonla fruit

p. 109

Custard Apple Orchard

p. 129

Custard Apple Fruit

p. 129

Tamarind Tree

p. 142

Fig Orchards

p. 157

Fig Fruits

p. 158

p. 179